THE CNC WORKSHOP

A MULTIMEDIA INTRODUCTION TO COMPUTER NUMERICAL CONTROL

VERSION 2.0

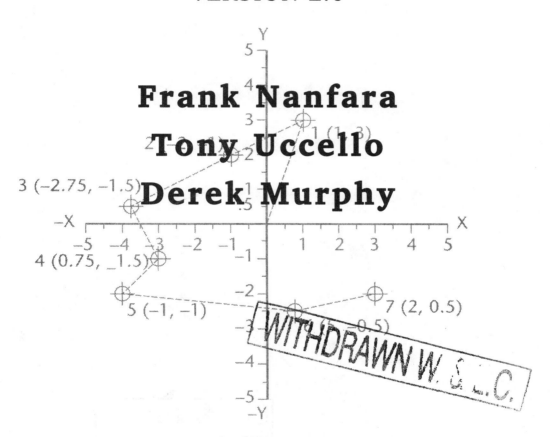

Frank Nanfara
Tony Uccello
Derek Murphy

SDC
PUBLICATIONS

Schroff Development Corporation

WWW.SCHROFF.COM

ISBN: 978-1-58503-083-5

Printed in the United States of America

LICENSE AGREEMENT AND LIMITED WARRANTY

READ THE FOLLOWING TERMS AND CONDITIONS CAREFULLY BEFORE OPENING THIS SOFTWARE PACKAGE. THIS LEGAL DOCUMENT IS AN AGREEMENT BETWEEN YOU AND TORCOMP SYSTEMS LTD. (THE "COMPANY"). BY OPENING THIS SEALED SOFTWARE PACKAGE YOU ARE AGREEING TO BE BOUND BY THESE TERMS AND CONDITIONS. IF YOU DO NOT AGREE WITH THESE TERMS AND CONDITIONS, DO NOT OPEN THE SOFTWARE PACKAGE. PROMPTLY RETURN THE UNOPENED SOFTWARE PACKAGE AND ALL ACCOMPANYING ITEMS TO THE PLACE YOU OBTAINED THEM FOR A FULL REFUND OF ANY SUMS YOU HAVE PAID.

1.   **GRANT OF LICENSE:** In consideration of your purchase of this book, and your agreement to abide by the terms and conditions of this Agreement, the Company grant to you a nonexclusive right to use and display the copy of the enclosed software program (hereinafter the "SOFTWARE") on a single computer (i.e., with a single CPU) at a single location so long as you comply with the terms of this Agreement. The Company reserves all rights not expressly granted to you under this Agreement.

2.   **OWNERSHIP OF SOFTWARE:** You own only the magnetic or physical media (the enclosed media) on which the SOFTWARE is recorded or fixed, but the Company and the software developers retain all the rights, title, and ownership to the SOFTWARE recorded on the original media copy(ies) and all subsequent copies of the SOFTWARE, regardless of the form or media on which the original or other copies may exist. This license is not a sale of the original SOFTWARE or any copy to you.

3.   **COPY RESTRICTIONS:** This SOFTWARE and the accompanying printed materials and user manual (the "Documentation") are the subject of copyright. The individual programs on the media are copyrighted by the authors of each program. Some of the programs on the media include separate licensing agreements. If you intend to use one of these programs, you must read and follow its accompanying license agreement.
You may NOT copy the Documentation or the SOFTWARE, except that you may make a single copy of the SOFTWARE for backup or archival purposes only. You may be held legally responsible for any copying and copyright infringement, which is caused or encouraged by your failure to abide by the terms of this restriction.

4.   **USE RESTRICTIONS:** You may NOT network the SOFTWARE or otherwise use it on more than one computer or computer workstations at the same time unless a network license has been granted to you separately. You may physically transfer the SOFTWARE from one computer to another provided that the SOFTWARE is used on only one computer at a time. You may NOT distribute copies of the SOFTWARE or Documentation to others. You may NOT reverse engineer, disassemble, decompile, modify, adapt, translate, or create derivative works based on the SOFTWARE or Documentation without prior written consent of the Company. You may NOT rent or lease this SOFTWARE.

5.   **TRANSFER RESTRICTIONS:** The enclosed SOFTWARE is licensed only to you and may NOT be transferred to any one else without the prior written consent of the Company. Any unauthorized transfer of the SOFTWARE shall result in the immediate termination of this Agreement.

6.   **TERMINATION:** This license is effective until terminated. This license will terminate automatically without notice from the Company and become null and void if you fail to comply with any provisions or limitations of this license. Upon termination, you shall destroy the Documentation and all copies of the SOFTWARE. All provisions of this Agreement as to warranties, limitations of liability, remedies or damages, and our ownership rights shall survive termination.

7.   **MISCELLANEOUS:** This Agreement shall be construed in accordance with the laws of the United States of America and the State of Kansas and shall benefit the Company, its affiliates, and assignees.

8.   **LIMITED WARRANTY AND DISCLAIMER OF WARRANTY:** The Company warrants that the SOFTWARE, when properly used in accordance with the Documentation, will operate in substantial conformity with the description of the SOFTWARE set forth in the Documentation. The Company does NOT warrant that the SOFTWARE will meet your requirements or that the operation of the SOFTWARE will be uninterrupted or error-free. The Company warrants that the media on which the SOFTWARE is delivered shall be free from defect in materials and workmanship under normal use for a period of thirty (30) days from the date of your purchase. Your only remedy and the Company's only obligation under these limited warranties is, at the Company's option, return of the warranted item for a refund of any amounts paid by you or replacement of the item. Any replacement of SOFTWARE or media under the warranties shall not extend the original warranty period. The limited warranty set forth above shall NOT apply to any SOFTWARE which the Company determines in good faith has been subject to misuse, neglect, improper installation, repair, alteration, or damage by you. EXCEPT FOR THE EXPRESSED WARRANTIES SET FORTH ABOVE, THE COMPAY DISCLAIMS ALL WARRANTIES, EXPRESS OR IMPLIED, INCLUDING WITHOUT LIMITATION, THE IMPLIED WARRANTIES OF MERCHANTABILITY AND FITNESS FOR A PARTICULAR PURPOSE, EXCEPT FOR THE EXPRESS WARRANTY SET FORTH ABOVE, THE COMPANY DOES NOT WARRANT, GUARANTEE, OR MAKE ANYY REPRESENTATION REGARDING THE USE OR THE RESULTS OF THE USE OF THE SOFTWARE IN TERMS OF ITS CORRECTNESS, ACCURACY, RELIABILITY, CURRENTNESS, OR OTHERWISE.

IN NO EVENT, SHALL THE COMPANY OR ITS EMPLOYEES, AGENTS, SUPPLIERS, OR CONTRACTORS BE LIABLE FOR ANY INCIDENTAL, INDIRECT, SPECIAL, OR CONSEQUENTIAL DAMAGES ARISING OUT OF OR IN CONNECTION WITH THE LICENSE GRANTED UNDER THIS AGREEMENT, OR FOR LOSS OF USE, LOSS OF DATA, LOSS OF INCOME OR PROFIT, OR OTHER LOSSES, SUSTAINED AS A RESULT OF INJURY TO ANY PERSON, OR LOSS OF OR DAMAGE TO PROPERTY, OR CLAIMS OF THIRD PARTIES, EVEN IF THE COMPANY OR AN AUTHORIZED REPRESENTATIVE OF THE COMPAY HAS BEEN ADVISED OF THE POSSIBILITY OF SUCH DAMAGES. IN NO EVENT SHALL LIABILITY OF THE COMPANY FOR DAMAGES WITH RESPECT TO THE SOFTWARE EXCEED THE AMOUNTS ACTUALLY PAID BY YOU, IF ANY, FOR THE SOFTWARE.

SOME JURISDICTIONS DO NOT ALLOW THE LIMITATION OF IMPLIED WARRANTIES OR LIABILITY FOR INCIDENTAL, INDIRECT, SPECIAL, OR CONSEQUENTIAL DAMAGES, SO THE ABOVE LIMITATIONS MAY NOT ALWAYS APPLY. THE WARRANTIES IN THIS AGREEMENT GIVE YOU SPECIFIC LEGAL RIGHTS AND YOU MAY ALSO HAVE OTHER RIGHTS WHICH VARY IN ACCORDANCE WITH LOCAL LAW.

ACKNOWLEDGEMENT

YOU ACKNOWLEDGE THAT YOU HAVE READ THIS AGREEMENT, UNDERSTAND IT, AND AGREE TO BE BOUND BY ITS TERMS AND CONDITIONS. YOU ALSO AGREE THAT THIS AGREEMENT IS THE COMPLETE AND EXCLUSIVE STATEMENT OF THE AGREEMENT BETWEEN YOU AND THE COMPANY AND SUPERSEDES ALL PROPOSALS OR PRIOR AGREEMENTS, ORAL, OR WRITTEN, AND ANY OTHER COMMUNICATIONS BETWEEN YOU AND THE COMPANY OR ANY REPRESENTATIVE OF THE COMPANY RELATING TO THE SUBJECT MATTER OF THIS AGREEMENT.

PREFACE

Computer numerical control (CNC) programming skills are important to any engineering student. *The CNC Workshop: An Multimedia Introduction to Computer Numerical Control*, an affordable text, software, and CBT combination, is designed to introduce students to CNC and provide them with an opportunity to learn and practice at their own pace. Though technology has been constantly refining the CNC manufacturing process, the basic principles remain the same. This latest text and CD-ROM combination utilizes many new Internet technologies to deliver a rich learning experience to the student. The emergence of Java as an Internet programming language has enabled TORCOMP, the developers, to bring this exciting new technology to the forefront. TORCOMP's CNCez Java Edition is the first CNC simulation software to run both in a Web Browser environment and as a standalone Windows Java application. The workbook teaches students about CNC programming. The CBT and software, CNC Workshop with TORCOMP CNCez Java Edition, enforces important concepts and allows students to create CNC programs and test them in a simulated manufacturing environment.

The CNC Workshop may be used as a complete text for a course in computer numerical control or as an instructional supplement if CNC is a component of another course. Students could use this package to learn about CNC on their own outside of the classroom. The software will run on current PCs with the latest Windows operating systems and Internet Explorer and is ideal for learning, teaching, and testing generic CNC programming.

FEATURES OF THE WORKBOOK

The text introduces students to CNC programming and the use of the TORCOMP CNCez Java Edition simulation software, providing students with the basics they need to gain confidence in CNC programming.

- Chapters 1 through 3 discuss the basic concepts of computer numerical control, including its theoretical and applied aspects.

- Chapter 4 introduces students to the interface of the TORCOMP CNCez Java Edition simulation software programs.

- Chapters 5 and 6 begin teaching students to program and run simulations. Chapter 5 describes milling, and Chapter 6 covers turning, the two most important and most common uses for CNC programming. Each chapter concludes with five step-by-step tutorials to help the student develop programming skills.

- Chapter 7 describes the basics of computer-aided design and computer-aided manufacturing with simple examples of creating part geometry and generating CNC code.

- Chapter 8 provides additional exercises, including questions to answer and new advanced simulations to program.

FEATURES OF THE CBT

The use of multimedia technologies is intented to motivate the student to develop the discipline that is necessary to learn CNC programming. Students are kept involved in the content by means of interactive demonstrations and hands-on examples. Navigation is simple; students can move freely among content, help topics, simulators, and glossaries. Several galleries also are included, which showcase industrial applications, machines in manufacturing settings, and interactive virtual reality modeling language (VRML) model examples.

FEATURES OF THE SIMULATION SOFTWARE

One of the biggest problems in learning CNC technology and specifically CNC programming is finding time for the hands-on training required on a CNC machine tool. *The CNC Workshop* lessens the dependence on traditional classroom-style learning and maximizes the amount of hands-on CNC programming time available to each student. Using the simulation software allows the student to perform dry runs of all CNC programs in a simulated manufacturing environment. TORCOMP's CNCez Java Edition allows students to define and save their custom machine configurations, as well as the ability to display machined solid models of their programs. By using TORCOMP CNCez Java Edition to test a program, students need to use a CNC machine tool only for final confirmation of a design.

Two other key features of this learning system are safety and comfort. Because this software emulates a CNC milling machine or CNC lathe, the students can do all the necessary CNC programming training and later write, edit, and test new CNC programs on the computer. Because a CNC machine is not actually being used, learning can take place in a safe and comfortable environment. Another special feature of the TORCOMP CNCez Java Edition is that the on-screen simulator and the programming editor are interactive. As the student creates and edits CNC programs, the screen graphics immediately show the result of each step, providing instant feedback. Students will know immediately which part of their program generated which result.

For users with prior knowledge of and experience with CNC programming, the TORCOMP CNCez Java Edition simulation software can be used to test existing generic CNC programs, which follow the EIA/ISO standards, and to write and edit new ones. The workbook or CD-ROM can be used as a reference guide, or the user may select the Help menu option for assistance. Please read the readme.doc file on the CD for the latest on any changes to this text.

A WARNING TO THE READER

The objective of this workbook, CD-ROM and software combination is to teach CNC programming. It is *not* to instruct the user on machining terms or procedures, nor does skill in the use of the simulation software mean that hands-on experience with manufacturing equipment can be eliminated. All speeds and feeds used in this manual have been tested on machinable wax or plastics only and are not recommended for harder materials or metals.

All programs in this manual have been tested for reliability. Some changes may be required to adapt these programs to non-TORCOMP CNC machines. Please consult your CNC machine operations manual.

ACKNOWLEDGMENTS

We would like to thank our development team, especially Patrick Bourke, Peter Li, and Thanh Vo for their outstanding Java efforts; Diem Trang Nguyen for truly applying herself in researching 3D and OpenGL; Domenic Grande for his foresight in modern graphic design techniques to both the workbook and the software. We appreciate the assistance of Gregory Arceri from the Ford Design Center, John Arnone from Ford of Canada, Henry Johnson from Johnson Precision Products Ltd., Dave Yared of Carbaloy Inc., USA, Michael Yeates of the MIT Musem, Dave Baldwin and Alan Shoemaker from the Mitre Corporation, and everyone else who contributed various materials from still photos to videos to historic film footage. We would like to thank all the people who reviewed the manuscript and CD-ROM for their comments and suggestions, especially Denise Olson, Phoebe Ling for having faith in us. And special thanks go to our families for being so patient and helping us see this project come to light.

Frank Nanfara
Tony Uccello
Derek Murphy

CONTENTS

M-Codes 109

CHAPTER 1

Introduction to CNC

After studying this chapter, the student should have knowledge of the following:

> The evolution of CNC
>
> The process of CNC
>
> The flow of CNC processing
>
> The objectives of CNC

INTRODUCTION

Computer numerical control (CNC) is the process of manufacturing machined parts. Production is controlled and allocated by a computerized controller. The controller uses motors to drive each axis of a machine tool and actually regulates the direction, speed, and length of time each motor rotates. A programmed path is loaded into the machine's computer by the operator and then executed. The program consists of numeric point data in conjuction with specialized machine control commands and function codes. Numerical control (NC) is the original term given to this technology and is still often used interchangeably with CNC.

NC technology has been one of manufacturing's major developments in the past 50 years. It not only resulted in the development of new techniques and the achievement of higher production levels, but it also helped increase product quality and stabilize manufacturing costs.

THE EVOLUTION OF NC

The principal of NC manufacturing has been evolving since the Industrial Revolution, although the actual procedures involved have developed with technology. Early attempts to automate production made use of belts, pulleys, and cams. However, manual labor was by far more cost effective than were the development and operation of big, new machines.

Not until World War II did industrialists realize that they couldn't meet both quantity and quality requirements at the same time. Machinists of the day could produce superior quality parts but not at high volumes. As the quantity of a certain product increased, the quality decreased due to the human factors involved.

During World War II the United States Army Ballistic Research Lab and the University of Pennsylvania collaborated on development of ENIAC, the world's first digital computer. This was an extremely large vacuum tube computer, which was used to calculate ballistic trajectory tables for artillery. Programming involved setting hundreds of switches and cables manually prior to having the machine sequence through the settings. In

The effects of World War II were significant in the development of NC machinery. (Courtesy U.S. Air Force.)

The world's first digital computer, ENIAC, contained more than 30,000 vacuum tubes and miles of cable. (Courtesy U.S. Army.)

the early 1950s the Massachusetts Institute of Technology developed a more advanced vacuum tube computer called the Whirlwind. This improved computational device was capable of executing thousands of times more instructions per second than ENIAC.

To ensure that all U.S. military airplanes were manufactured identically, the United States Air Force invited several companies to develop and manufacture numerical control systems that could handle the volume and repeatability.

The specific goals of developing NC were to:

1. Increase production
2. Improve the quality and accuracy of manufactured parts
3. Stabilize manufacturing costs
4. Manufacture and assemble complex parts quickly

IMPORTANT

Along with programmable automation, NC was designed to readily accommodate changes in product design and to help produce parts that:

- Were similar in terms of the raw materials used
- Varied in size and geometry
- Were made in small- to medium-sized batches
- Required a sequence of similar steps to complete each workpiece

The first contract was awarded to the Parsons Corporation of Michigan, which had developed a control system that directed a spindle to many points in succession. Although the contract date was June 15, 1949, the demand for these systems was a direct result of the war effort.

In 1951, the Servomechanism Laboratory of the Massachusetts Institute of Technology (MIT) was given a subcontract by Parsons to develop a servo system for the machine tool. As MIT was also working on the Whirlwind at the time, the total NC development project thus was conducted at MIT.

In 1952, the first three-axis, numerically controlled, tape-fed machine tool was created. A Cincinnati Milacron Hydro-Tel Vertical Spindle

The Servo Mechanism Laboratory at MIT. (Courtesy MIT Musem Library.)

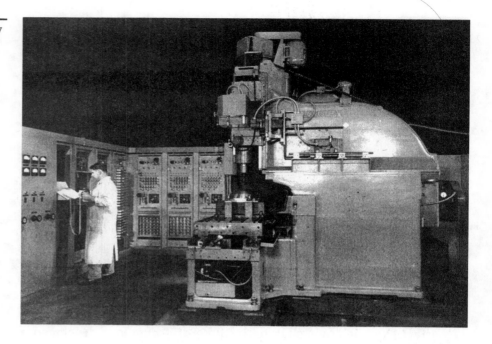

milling machine was retrofitted and controlled by the Whirlwind computer. The controller was equipped with optical sensors and used a straight binary perforated tape to hold the instructions; the tape was read via a mechanical feeding mechanism. In 1954, numerical control was announced to the public, and three years later the first production NC machines were delivered and installed.

By 1960, NC was widely accepted and readily available. Although the controllers used alphanumeric characters in their controlling code, it was still called numerical control.

The majority of these first generation NC machine tools required coded paper tape to run. The engineers would generate NC code on their computers, then encode long strips of paper tape by punching hole patterns in them. Because these hole patterns were encoded, it was difficult to identify part programs easily. Later, the use of "man-readables" solved

The first NC machine producing identical parts. (Courtesy MIT Musem Library.)

An early NC paper tape reader.
(Courtesy Giddings and Lewis.)

this problem. Man-readables were nothing more than a special hole
punching technique that yielded alphanumeric hole patterns that could
be easily read by the operator. These were usually punched on a leader
before the actual program started. Sometimes these punched tapes were
very long and had to be stored on large reels. These reels of tape would
then be taken to the shop floor, where the operator would insert them into
the controller and run them. The tapes contained all the data required for
machine operation. The machine tool would then perform exactly the
same operations as many times as required.

Further research and development brought about new generations of
NC machines. The subsequent introduction of computer numerical control, whereby a computer is used to control the machine tool, eliminated
the dependence on bulky and fragile paper tape.

Direct numerical control and distributed numerical control (DNC) describe communications to a machine tool from a remote computer, as
shown in Fig. 1.1. In direct numerical control, part program instruction
blocks are communicated to a machine tool as required and as fast as the
machine can accept them. This method of communication was very popular because it eliminated the paper tape system and increased the maximum length of a program. In distributed numerical control, whole
programs or multiple programs are communicated to a CNC machine tool
or several CNC machine tools, usually via an RS232 serial communications link. This process was made possible by the increased memory

A paper tape leader showing man-
readables.

PC DNC system

PC

Ethernet or token ring network

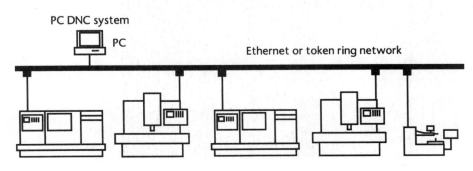

FIGURE 1.1
A diagram of a modern DNC
layout.

capacity of CNC controllers. In the early days, controllers could not store programs, so programs were stored on paper tape. Later controllers were able to store only limited-sized programs. Modern controllers can store hundreds of programs on their built-in hard drive memory systems. With today's increased widespread use of networks, part programs can be catalogued on a central server computer and any CNC machine on the factory network, either locally or even remotely, utilizing popular Ethernet or TCP/IP Internet protocols, can request a specific CNC program file instantaneously even thousands of miles away. This modern setup centralizes the design and engineering center away from the shop floor. It also eliminates the need to have the part programs stored in the same location as the machine tools. This is also beneficial in improved production scheduling.

From 1955 to 1960, MIT also developed a computer-assisted programming system called automatically programmed tools (APT). This programming language was developed to ease the task of complicated three-axis programming, mainly in the aerospace industry. APT uses English-like words to describe the geometry and the tool motions on a part program. Figure 1.2 shows the point, line, and circle specification commands along with the movement commands that follow.

The great advantage of early modern CNC was the ability of the code-generating computer to move from the engineering department back onto the shop floor, where it was directed by the machinist. The computer and machine control unit (MCU) were now one unit, capable of creating the program and then storing it in memory and running it on demand. This also eliminated much of the need for paper tape. With the exception of

FIGURE 1.2
Example of an APT program.

```
PARTNO 3764
MACHIN/2167
CUTTER/.375
FEDRAT/5
SP = POINT/.25,.25,.5
P1 = POINT/0,0,.5
P2 = POINT/.25,.25,-.125
P3 = POINT/.5,.25,-.125
L1 = LINE/P2, ATANGL,0
C1 = CIRCLE/(1.25 +1.75),.375,.375
C2 = CIRCLE/1.750,1.950, .5
L2 = LINE/RIGHT, TANTO, C1, RIGHT, TANTO, C2
L3 = LINE/P1, LEFT, TANTO, C2
FROM/SP
GO/TO, P1
GO/TO, P2
GORGHT/L2,TANTO C1
GORGHT/L3, TANTO C2
GO/TO, SP
FINI
```

extremely large programs, most NC programs are stored on the machine tool's memory built into a computerized controller.

Machining centers, which take the place of half a dozen machines, are now capable of many operations—including milling, boring, drilling, facing, spotting, counterboring, threading, and tapping—all in one setup. These machines are mainly used in mid- to large-sized production runs and sometimes are dedicated to a family of parts manufacturing.

Now a well-established technology, CNC machines have become commonplace. Over 85 percent of machine tools manufactured today are CNC. The number of manufacturing systems has blossomed, and most engineering companies utilize high levels of computerization.

Computer integration can be implemented at almost any level—from a simple machine shop with a simple computer-aided manufacturing (CAM) system, to companies with several dispersed engineering, design, and production sites and many hundreds of machines and systems.

CNC technology has the following advantages over NC technology: IMPORTANT

1. Programs can be entered at the machine and stored in memory.
2. Programs are easier to edit, so part programming and process design time are reduced.
3. There is greater flexibility in the complexity of parts that can be produced.
4. Three-dimensional geometric models of parts, stored in the computer, can be used to generate CNC part programs almost automatically, thus saving manual programming time.
5. Computers can be connected to other computers worldwide, either by direct modem connection or through a network, thereby allowing part programs to be transmitted directly to remote CNC machines.

CNC technology has the following disadvantages:

1. CNC is slightly more expensive, although today it would be rare to find an NC machine tool sold that is not CNC.
2. Possibly more training is required for the machine operator. This, however, depends on the complexity of the machine tool, as a CNC machine may actually require less training. Modern CNC controllers are now very user-friendly. This shortens the learning curve for machine operators, making the cost recovery period much shorter.
3. Maintenance costs may be greater.

MICROCOMPUTER TECHNOLOGY

The modern computer is an electronic machine that performs mathematical or logical calculations in accordance with a predetermined set of instructions. The computer itself is called the hardware; the programs that run on the computer are called software.

The three basic components of a computer, as shown in Fig. 1.3, are: REMEMBER

1. Central processing unit (CPU)

FIGURE 1.3
A diagram of the basic components that make up a computer.

```
                         ┌──────────┐
                         │  Output  │
                         └────┬─────┘
   ┌──────────┐   ┌──────────┴┐   ┌──────────┐
   │  Memory  │───│    CPU    │───│ Storage  │
   └──────────┘   └─────┬─────┘   └──────────┘
                        │
                   ┌────┴─────┐
                   │  Input   │
                   └──────────┘
```

2. Memory

3. Input/output section

The CPU controls and sequences the activities of the computer components and performs the various arithmetic and logical operations. The memory is used by the CPU to store, retrieve, and manipulate data. The input/output section interprets incoming and outgoing signals that direct the CPU's operations.

Various peripheral devices associated with computers include monitors, scanners, printers, and plotters.

The use of computers in industry is now commonplace. Cheaper and faster personal computers have allowed companies to introduce computers at all levels. Various forms of computer-assisted programming are available for both shop floor control and NC programming.

The CNCez Solid View displays the results of the CNC program being executed. This is extremely useful for tool proving prior to running the program on the machine tool.

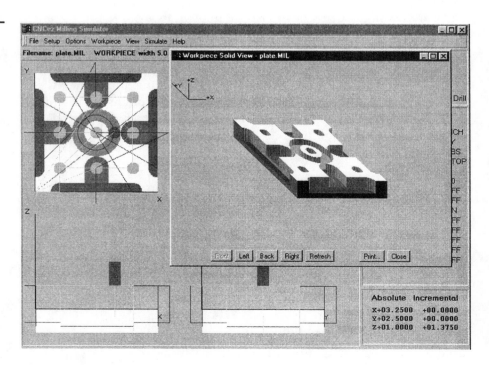

COMPUTER BASICS

Although computers have enhanced NC to a great extent, certain rules still remain in effect. One rule that still holds but is easier to live with today is:

Garbage in = Garbage out.

This rule, common in industry, was truer in the past than it is today. When writing NC programs using the old systems, programmers could check for errors only by running the program on the machine tool it was targeted for. Incorrect data entered resulted in the machine not running as required or not running at all—therefore the saying. With today's more powerful computers and software, such as CNCez, syntax checking and simulation enable the programmer to verify programs without running them on a production CNC machine tool, hence lowering production time and overall costs, and freeing up machine tools for manufacturing.

NC APPLICATIONS

From the days of the first NC milling machine, there have been many applications for NC technology, ranging from milling, turning, and electric discharge machining (EDM) to laser, flame and plasma cutting, punching and nibbling, forming, bending, grinding, inspection, and robotics.

Although aerospace is still one of the principal industries that require and use NC technology extensively, other industries have also embraced it. Because of the continuing advances in computers and their affordability, the cost of NC technology has been dropping rapidly. Now, even small machine shops and small specialty industries have come to acquire this technology.

Today you can find NC products in many areas ranging from metalworking and automotive to electronics, appliances, engraving, sign making, jewelery design, and furniture manufacturing.

MILLING

The process of milling involves the use of a rotating cutter to remove material from a workpiece. Single- or multiple-axis control moves can generate either simple two-dimensional patterns or profiles, or complex three-dimensional shapes.

TURNING

Turning utilizes a cutter that moves perpendicularly through the center plane of a rotating workpiece. The part shape depends on the shape of the tool and the operations performed to obtain the finished part.

WIRE EDM

Electrical discharge machining, or wire EDM, uses an electrical discharge from a thin wire to achieve fine cuts through hard metal parts. Most EDM machines use two parallel planes in which each cutting point can move independently of the other. This is useful in producing tapered pieces used in the production of punch dies for stamping.

A Cincinnati Milacron Arrow 750
Machining Center with an A-Axis
rotary option.
(Courtesy of Cincinnati Milacron.)

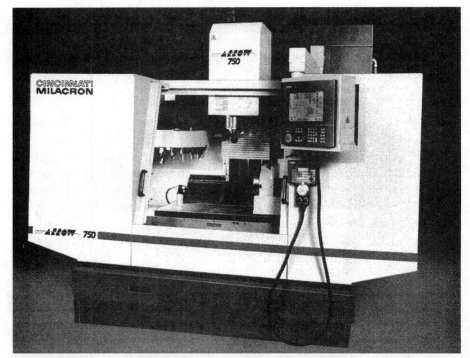

A Cincinnati Milacron Universal
5-Axis Machining Center with
automatic pallet changer and chip
conveyor option.
(Courtesy of Cincinnati Milacron.)

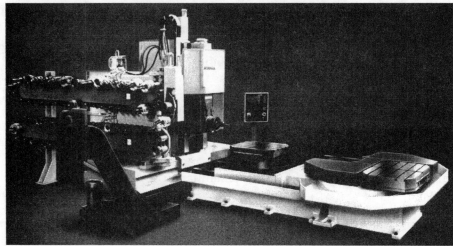

A large Cincinnati Milacron
5-Axis Gantry Profiling Mill.
(Courtesy of Cincinnati Milacron.)

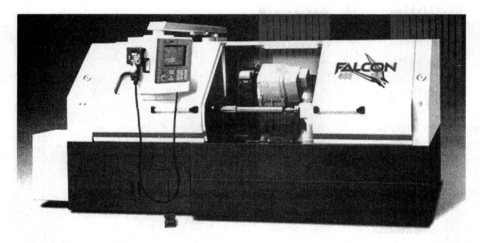

A Cincinnati Milacron Falcon
Turning Center.
(Courtesy Cincinnati Milacron.)

LASER, FLAME, AND PLASMA

Laser, flame, and plasma cutting use a powerful beam of light, a concentrated flame, or a plasma arc, respectively, to remove material. Depending on the target and thickness of the material, each application has certain advantages.

PUNCHING AND NIBBLING

Punching and nibbling are used to cut patterns in sheets of metal by the use of punch dies. Repeated punches along a path achieves a nibbling effect that allows cutting of complex patterns, which would otherwise be very difficult with conventional means. Forming and louvering are also typical applications of CNC punch machines.

ROBOTS AND CNC

The widespread use of CNC in manufacturing is ideal for the use of industrial robots to perform repetitive tasks. Such tasks may involve handling heavy and sometimes hazardous materials. Sophisticated CNC machining centers can contain pallet changers and special interfaces

Strippit Fabri-Center punching
and nibbling machine.
(Courtesy of Strippit, Inc.)

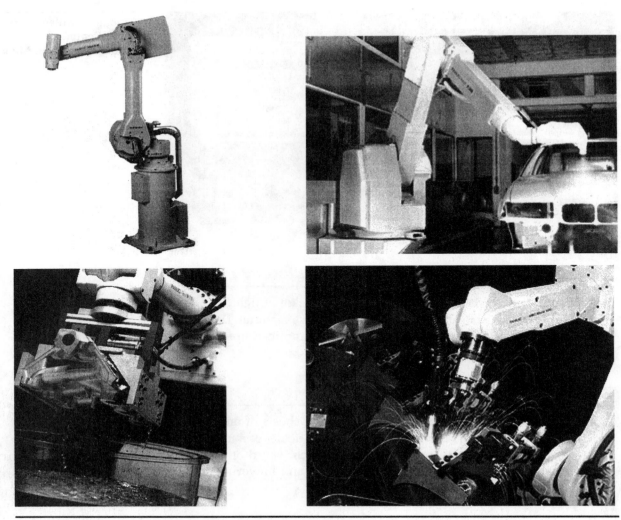

Fanuc industrial robots used in various applications. (Courtesy of Fanuc Robotics.)

that can easily accommodate industrial robots. Such specialty operations are commonplace in today's high-volume machining environments.

Although standards exist and there are similarities among controllers, many differences exist among the various controllers on the market. One manufacturer of controllers may have more than a dozen different models and a dozen different variations of one model alone. Although the basic principal of controllers is easy to understand, the thousands of variations of machine tools and applications require many different types of controllers. It is common to find two different machine tools using the same model of controller with slightly different options.

CONTROLLER STANDARDS

To understand CNC, you must first understand both the differences and similarities of controllers on the market. In addition to the differences in controllers based on the variety of machine tools and applications, other differences relate to the manufacturers and the standards, if any, they follow and how closely they follow those standards.

Fanuc Mfg. cell with several robots and CNC machines. (Courtesy of Fanuc Robotics.)

GE Fanuc 18-T CNC control used on a Hardinge® CONQUEST® T42 CNC Turning Center. Hardinge and CONQUEST are registered trademarks of Hardinge Brothers, Inc.

The TORCOMP Desktop CNC Gantry Training Machine, showing the components of a typical CNC machine tool.

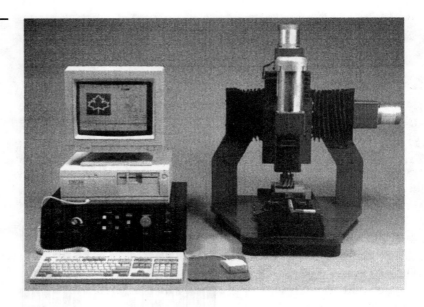

CNC CONTROLLERS

IMPORTANT

There are three major components of a CNC machine tool:

1. The machine tool itself, which can be any one of many different types.
2. The motors and feedback mechanisms, which are very important because they are the link between the machine tool and the controller. Therefore the type, size, and resolution are very important considerations for different applications.
3. The heart of the CNC machine tool—the controller or machine control unit (MCU).

EIA AND ISO STANDARDS

The International Standardization Organization (ISO) and the Electronic Industries Association (EIA) developed two very similar standards that are generally followed worldwide: the ISO 6983 and the EIA RS274. Some countries may have their own standards, but most follow ISO and EIA. The main standards for NC provide simple programming instructions to enable a machine tool to carry out a particular operation. For example, the following lines of code will instruct a CNC milling machine that, on executing line or block number 100, the tool is to cut relative to the origin point at a feedrate of 20 in./min along the X axis 1.25 in. and the Y axis 1.75 in.

```
N95 G90 G20
N100 G01 X1.25 Y1.75 F20.0
```

Axis designation on a machine tool and the coordinate system, both of which are covered in more detail in Chapter 2, are also standardized by EIA 267-C. This standard applies to all NC machine tools regardless of whether the controller follows a particular standard. This standard is equally important, if not more so, than the EIA RS274 standard, as it forms the link to computer-aided design and computer-aided manufacturing (CAD/CAM), which follow similar standards.

CONVERSATIONAL (NONSTANDARD)

Though rarer than conventional CNC controllers, an alternative is the conversational CNC controller. These controllers generally do not follow any standard, are mostly proprietary, and are supposed to be easy to use. The operator does not need to know how to program but only how to read and respond to the prompts on the controller screen. Generally, whereas simple machines that produce simple parts may use this system, more complex machines producing more complex parts may not. Therefore some CNC machines may offer both ISO/EIA standard programming and conversational programming. Besides the nonstandardization of conversational CNC controllers, their other drawback is that communication from CAD/CAM systems becomes more difficult. In general, a stand-alone machine, one that does not require programs created through a CAD/CAM system and will produce simple parts, is a good candidate for a conversational controller.

As you learn more about the CNC industry, you will soon discover that there are several main controller manufacturers. They include Fanuc, General Electric, Mitsubishi, Yasnak, and Bendix. However, some CNC machine tool companies, such as Cincinnati Milacron, Giddings, and Lewis and Bridgeport, may use their own proprietary controllers for their machine tools. In general, most of these companies follow the EIA/ISO program standards, so their programs are quite portable.

As you proceed through the rest of this workbook, the CD-ROM, and the CNCez simulation software, keep in mind that the standards used follow as closely as possible the EIA RS274 standard for basic three-axis NC Milling and two-axis NC Turning. Therefore programs developed with CNCez may require some modification for your particular machine tool. If you require more assistance, please consult the Technical Support section at the end of this workbook.

THE CNC PROCESS

In principal, the process of CNC manufacturing is the same as conventional manufacturing methods. Conventionally, shop drawings are generated by design engineers, who pass them to machinists. The machinists then read the drawings and methodically calculate toolpaths, cutter speeds, feeds, machining time, and the like.

In CNC programming, the machinist still has sole responsibility for the machine's operation. However, control is no longer exercised by manually turning the axis handwheels but through programming the use of the controller.

This is not to say that most proficient machinists will be computer programmers. Early CNC machines required manual input of G- and M-codes, but today a computer specialist is no longer needed for this task.

Note the following steps in CNC processing for both conventional and computer-aided methods.

IMPORTANT

IMPORTANT

FLOW OF CNC PROCESSING

1. Develop or obtain the part drawing.
2. Decide which machine(s) will perform the operations needed to produce the part.
3. Decide on the machining sequence and decide on cutter-path directions.
4. Choose the tooling required.
5. Do the required math calculations for the program coordinates.
6. Calculate the spindle speeds and feedrates required for the tooling and part material.
7. Write the CNC program.
8. Prepare setup sheets and tool lists (these will also be used for manufacturing operators).
9. Verify and edit the program, using either a virtual machine simulator such as CNCez or on the actual machine tool, creating a prototype.
10. Verify and edit the program on the actual machine and make changes to it if necessary.
11. Run the program and produce the final part.

IMPORTANT

FLOW OF COMPUTER-AIDED CNC PROCESSING

1. Develop or obtain the three-dimensional geometric model of the part, using CAD.
2. Decide which machining operations and cutter-path directions are required to produce the part (sometimes computer assisted or from engineering drawings and specifications).
3. Choose the tooling to be used (sometimes computer assisted).
4. Run a CAM software program to generate the CNC part program, including the setup sheets and list of tools.
5. Verify and edit the program, using a virtual machine simulator such as CNCez.
6. Download the part program(s) to the appropriate machine(s) over the network and machine the prototype. (Sometimes multiple machines will be used to fabricate a part.)
7. Verify the program(s) on the actual machine(s) and edit them if necessary.
8. Run the program and produce the part. If in a production environment, the production process can begin.

QUALITY CONTROL

As the operator uses the CNC machine tool and its controller, it will become evident which tools and machining procedures work best. This information should be documented, periodically reviewed, and used in all subsequent part programs for that particular machine. This becomes an iterative process, requiring constant improvement. Doing this will also enhance the efficiency of part programs and reduce runtime problems.

LAB EXERCISES

1. What is NC?

2. How did CNC come to be developed?

3. Draw a block diagram of a computer.

4. Draw a block diagram of a CNC mill.

5. Why are standards needed for CNC programming?

6. What is DNC?

7. List the steps in the CNC process.

8. Name some of the advantages of CNC.

9. What are some of the characteristics that CNC-produced parts should have?

10. Describe in your own words the CNC process.

CHAPTER 2

CNC Fundamentals and Vocabulary

After studying this chapter, the student should have knowledge of the following:

- The Cartesian coordinate system
- The motion directions of the CNC mill and lathe
- The types of coordinate systems
- Dimensioning theory
- The CNC vocabulary

AXIS AND MOTION NOMENCLATURE

All CNC machine tools follow the same standard for motion nomenclature and the same coordinate system. This, as mentioned in Chapter 1, is defined as the EIA 267-C standard. The standard defines a machine coordinate system and machine movements so that a programmer can describe machining operations without worrying about whether a tool approaches a workpiece or a workpiece approaches a tool.

Different machine tools have different machine motions, but they always use the same coordinate system. When describing a machine operation, the programmer always calculates tool movements relative to the coordinate system of the stationary workpiece. For example, you may have a CNC machine on which the tool is always stationary; however, the workpiece will move in various directions to achieve a finished part. In this example, when describing the tool motion or coordinate system, you describe the tool moving relative to the workpiece.

THE RIGHT-HAND RULE OF COORDINATES

The machine coordinate system is described by the right-hand rectangular coordinate system, that is, the rectangular Cartesian system. Based on this system, the right-hand rule governs how the primary axis of a machine tool should be designated.

As shown in Fig. 2.1, hold your right hand with the thumb, forefinger, and middle finger perpendicular to each other. The thumb represents the X axis, the forefinger the Y axis, and the middle finger the Z axis. The other two fingers are kept closed.

The direction of each finger represents the positive direction of motion.

The axis of the main spindle is always Z, and the positive direction is normally into the spindle.

REMEMBER

The right-hand rule is viewed from the programmer's perspective. For a mill the +Z axis always points into the spindle.

FIGURE 2.1
The right-hand rule of machine tool coordinates.

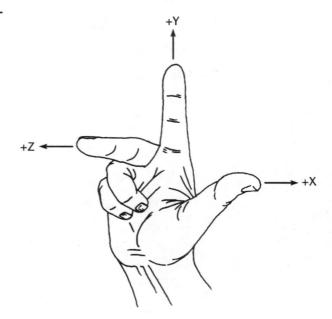

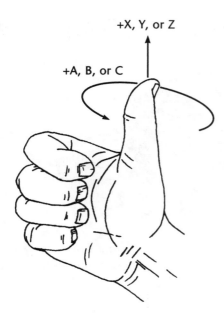

+X, Y, or Z

+A, B, or C

FIGURE 2.2
The right-hand rule for determining the clockwise rotary motion about X, Y, and Z.

On a mill the longest travel slide is usually designated as the X axis and is always perpendicular to the Z axis. On a lathe the longest travel slide is usually the Z axis.

If you rotate your hand—looking into your middle finger—the forefinger, which is perpendicular to it, represents the Y axis.

The base of your fingers are the start point, or origin (X0, Y0, Z0).

To determine the positive direction, clockwise, about an axis, close your hand with the thumb pointing out, as shown in Fig. 2.2, in the positive direction. The thumb may represent the X, Y, or Z axis direction, and the curl of the fingers may represent the clockwise, or positive, rotation about each axis. These are known as A, B, and C and represent the rotary motions about X, Y, and Z, respectively. These designations are used only for multiaxis machining centers. See Figs. 2.3–2.8 for examples of axis designations on various machine tools.

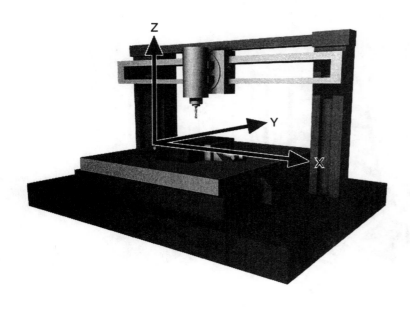

Z

Y

X

FIGURE 2.3
A typical three-axis CNC gantry milling machine.

FIGURE 2.4
A typical CNC lathe.

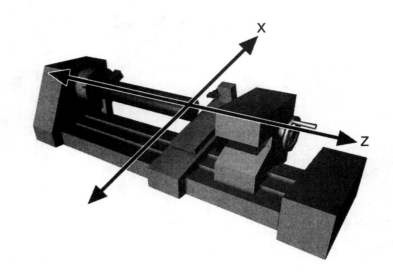

FIGURE 2.5
Axes orientation of a typical CNC knee mill.

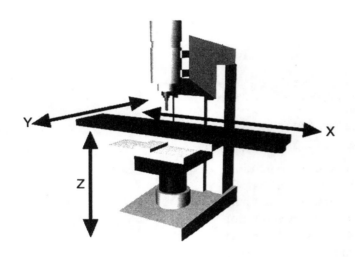

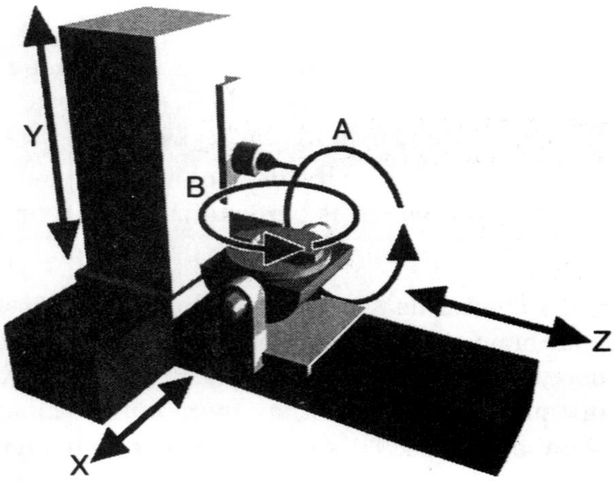

FIGURE 2.6
A horizontal milling machine with multiaxis rotary table.

FIGURE 2.7
A horizontal boring machine with rotating base and spindle head.

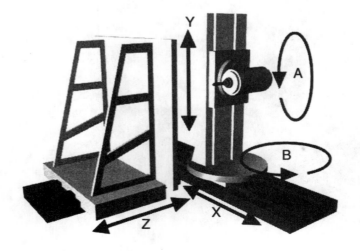

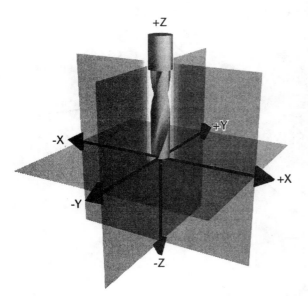

FIGURE 2.8
The Cartesian graph as it relates to the CNC machine tool. It shows the three axes, their planes, and the positive and negative directions the tool can move in.

CNC MILLING FUNDAMENTALS

THE CARTESIAN GRAPH FOR CNC MILLING

Note in Fig. 2.8 that the origin (reference point) is (X0,Y0,Z0). The direction of travel will dictate the value for the coordinate. Measuring the distance to a location from a fixed origin is referred to as the absolute coordinate system, and measuring the distance of a point relative to the last point is referred to as the incremental coordinate system. These systems are explained in more detail in the following sections.

Figure 2.9 shows the three planes in the Cartesian coordinate system: the XY plane, XZ plane, and YZ plane. The XY plane is the conventional standard.

There are two main reference points on a CNC machine from which to base all coordinates. The machine reference zero (MRZ) is a point on the actual machine. The part reference zero (PRZ) is a point on the actual part or workpiece.

All CNC machine tools require a reference point from which to base all coordinates. Although every CNC machine will usually have an MRZ, it is generally easier to use a point on the workpiece itself for reference. The reason is that the coordinates apply to the part anyway—hence the PRZ. It makes sense to put reference points on prominent objects, so the lower left-hand corner on top of the stock of each part is where the PRZ usually is defined. Figure 2.10 illustrates why the PRZ (or the origin point) is at the lower left-hand corner on top of the workpiece.

The advantages of having the PRZ at the lower left corner on top of the workpiece are as follows:

REMEMBER

1. Geometry creation is in the positive XY plane for CAD/CAM systems.

FIGURE 2.9
The three planes in the Cartesian coordinate system: XY, XZ, and YZ. The G-code notations are also displayed.

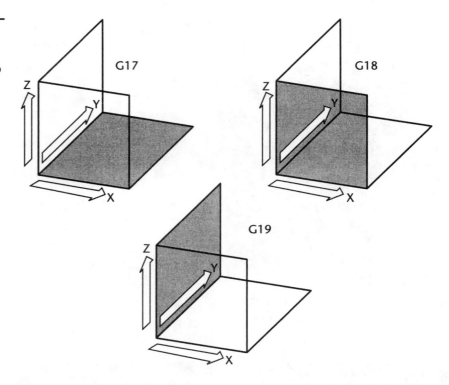

FIGURE 2.10
The PRZ is located at the lower left-hand corner on top of the workpiece to allow for easier coordinate measurement.

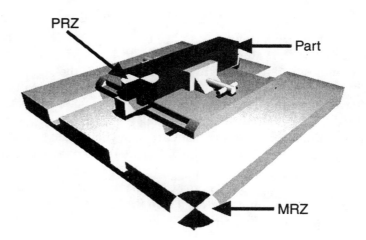

2. The corner of the workpiece is easy to find.
3. All negative Z depths are below the surface of the workpiece.

ABSOLUTE COORDINATES FOR MILLING

IMPORTANT

Absolute coordinates use the origin point as the reference point. This means that any point on the Cartesian graph can be plotted accurately by measuring the distance from the origin to the point, first in the X direction and then in the Y direction—then, (if applicable), in the Z direction. Points are generally written (X(+)(–)__, Y(+)(–)__, Z(+)(–)__), or, for example, (X3.25, Y–7.5, Z–0.5). Placing a positive sign before a number or a zero before a decimal point is usually optional.

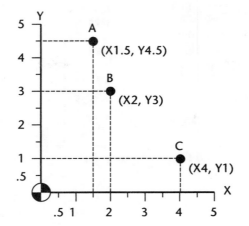

FIGURE 2.11
A Cartesian graph with points A, B, and C.

OBSERVE FROM FIG. 2.11 THE FOLLOWING.

Point A: This point is 1.5 units along the X axis from the origin and 4.5 units along the Y axis from the origin. It is at (X1.5, Y4.5).

Point B: This point is 2 units along the X axis and 3 units along the Y axis from the origin. It is at (X2.0, Y3.0).

Point C: Point C is 4 units along the X axis and 1 unit from the Y axis. It is at (X4.0, Y1.0).

With absolute coordinates, keep in mind that all coordinates are measured from (X0, Y0) to the point in question, first in the X direction, then in the Y direction, and finally in the Z direction. (Note that, if there is no Z coordinate, as on a two-dimensional part, it need not be included.)

The following examples, illustrated in Fig. 2.12, describe how the plus or minus values are derived. Remember, absolute coordinates are measured from the origin (0, 0) to the point, first in the X direction, then in the Y direction, and finally (if applicable) in the Z direction.

EXAMPLE A: From the origin, point A is 3 units along the +X axis and then down 2 units along the −Y axis. Therefore (X3.0, Y−2.0).

REMEMBER

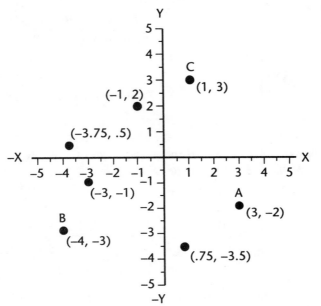

FIGURE 2.12
Examples of absolute coordinates. Note how the coordinates for each point are measured first in the X direction, then the Y direction to that point.

FIGURE 2.13
In this graph, point 1 is 1 unit in the +X direction from the origin and up 3 units in the +Y direction. Thus the incremental coordinates for point 1 are (X+1, Y+3). Point 2 is 2 units in the −X direction and down 1 unit in the −Y direction. Thus the incremental coordinates for point 2 are (X−2, Y−1).

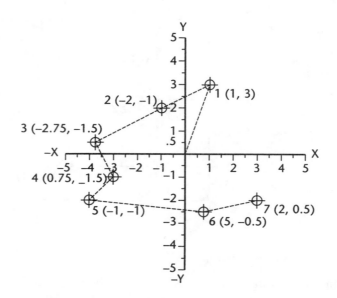

EXAMPLE B: From the origin, point B is 4 units along the −X axis and then down 3 units along the −Y axis. Therefore (X−4.0, Y−3.0).

EXAMPLE C: From the origin, point C is 1 unit along the +X axis and then up 3 units along the +Y axis. Therefore (X1.0, Y3.0).

INCREMENTAL COORDINATES FOR MILLING

IMPORTANT

Incremental coordinates use the present position as the reference point for the next movement. This means that any point in the Cartesian graph can be plotted accurately by measuring the distance between points, generally starting at the origin. It is important to remember that incremental coordinates are measured from point to point, always starting from a known reference point such as (0, 0).

> **EXAMPLE:** From Fig. 2.13, the incremental coordinates for points 3, 4, 5, 6, and 7 are as follows:
>
> Point 3 is (X−2.75, Y−1.5) units from the previous point (point 2).
> Point 4 is (X+.75, Y−1.5) units from the previous point (point 3).
> Point 5 is (X−1.0, Y−1.0) units from the previous point (point 4).
> Point 6 is (X+5, Y−.5) units from the previous point (point 5).
> Point 7 is (X+2.0, Y+.5) units from the previous point (point 6).

You generally use incremental coordinates when plotting a large series of points that are clustered around a reference point. In this way, you can use absolute coordinates to pinpoint the reference point (for example, a corner in a milled pocket or center of a bolt hole) and then use incremental coordinates to plot the points around it.

EXERCISES

To demonstrate an understanding of the characteristics and format of absolute and incremental coordinates, refer to Fig. 2.14 and complete the following exercises.

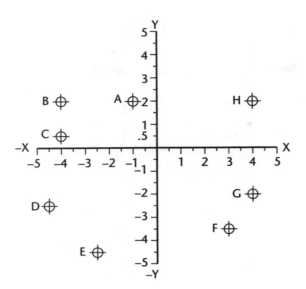

FIGURE 2.14
CNC milling coordinates.

EXERCISE 1: Absolute Coordinates

Fill in the X and Y blanks with the appropriate absolute coordinates for points A through H.

A: X_____, Y_____   B: X_____, Y_____

C: X_____, Y_____   D: X_____, Y_____

E: X_____, Y_____   F: X_____, Y_____

G: X_____, Y_____   H: X_____, Y_____

EXERCISE 2: Incremental Coordinates

Fill in the X and Y blanks with the appropriate incremental coordinates for points A through H.

A: X_____, Y_____   B: X_____, Y_____

C: X_____, Y_____   D: X_____, Y_____

E: X_____, Y_____   F: X_____, Y_____

G: X_____, Y_____   H: X_____, Y_____

CNC TURNING FUNDAMENTALS

All CNC lathes share the same two-axis coordinate system. This allows for the transfer of CNC programs among different machines, as all measurements are derived from the same reference points.

Basically, in CNC turning there is a primary (horizontal) axis and a secondary (vertical) axis. Figures 2.15–2.18 show that the primary axis is labeled Z and the secondary axis X.

It is also important to remember that on most CNC lathes, the tool post or turret is on the top, or back side, of the machine, unlike on a conventional lathe. This is why the tool is shown above the part.

FIGURE 2.15
The XZ Cartesian coordinate system. Note how the Z axis is horizontal and the X axis is vertical. The origin is at (X0 ,Z0), the intersection of the X and Z axes. Also note how each axis has a + and a − side. By taking the distance from (X0, Z0) and the direction (+ or −), you can accurately locate any point on this graph. Note that the origin is located on the right-hand side of the part to simplify programming moves.

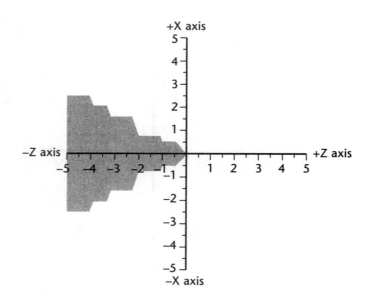

FIGURE 2.16
Relating the Cartesian graph to a CNC lathe. The major axis always runs through the spindle, so the Z axis is the longer one. The X axis is perpendicular to the Z axis.

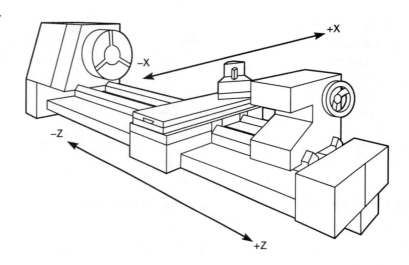

FIGURE 2.17
Merging the Cartesian graph with a lathe part. Note how the Z axis runs through the center of the part and the X axis is perpendicular to it. The origin is at the intersection of the X and Z axes at the center of the right-hand end of the workpiece.

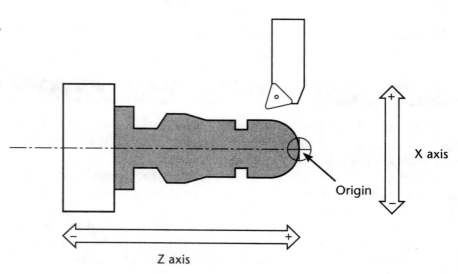

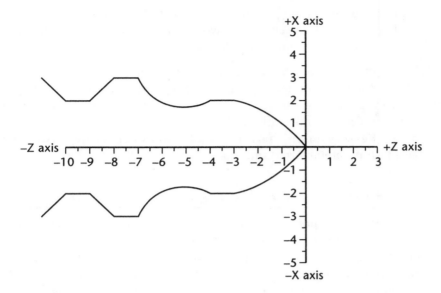

FIGURE 2.18
Note how the part sits on the graph: The origin is at the center of the right-hand end of the part. The Z axis mirrors the part, so only the top half of the part is required when programming. (We use the top half of the profile, because that is where the tool is.)

When locating points on a profile, you need not use the entire four-quadrant system. Any turned part is symmetric about the Z axis, so only its top half is required in a drawing. Compare Figs. 2.18 and 2.19 to see how the Cartesian graph is modified to suit the lathe application better.

When measuring X and Z coordinates, use a central reference point. Start all measurements at this reference point, the origin (X0, Z0). For our purposes, the origin is located at the center right-hand endpoint of the workpiece. Keep in mind that at times the center left-hand endpoint of the workpiece or even the chuck face may be used.

REMEMBER

DIAMETER VERSUS RADIUS PROGRAMMING

Diameter programming relates the X axis to the diameter of the workpiece. Therefore, if the workpiece has a 5-in. diameter and you want to command an absolute move to the outside, you would program X5.0.

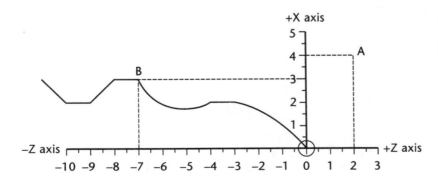

FIGURE 2.19
A typical lathe part drawing. Note how the whole profile fits in one quadrant and how all X values are positive and all Z values are negative.

Radius programming relates the X axis to the radius of the workpiece. Therefore, with the same size workpiece of 5 in., you would program X2.5 to move the tool to the outside.

Although many controllers can work in either mode, diameter programming is the most common and is the default with CNCez. To change to radius programming, select the Diameter Programming option of the Machine Configuration dialog from the Preferences... option of the File menu. **Keep in mind that all samples and step-by-step examples are based on diameter programming.**

ABSOLUTE COORDINATES FOR TURNING

IMPORTANT

When measuring points on a profile, you will usually find it easier to relate each point to the origin. Coordinates found in this way are called absolute coordinates because all values are absolute distances from the origin. The following section explains how to find points using absolute coordinates for both the radius and the diameter of the workpiece (see Fig. 2.20).

FINDING ABSOLUTE COORDINATES

When plotting points using absolute coordinates, always start at the origin (X0, Z0). Then travel along the Z axis until you reach a point directly below the point that you are trying to plot. Write down the Z value, then go up until you reach your point. Write down the X value. You now have the XZ coordinate for that point. Remember, travel left or right first along the Z axis and then up or down the X axis.

EXAMPLE A: Find point A.

1. Start at (X0, Z0).
2. Travel right until you are below point A.
3. Move up to point A.

The radial XZ coordinates for point A are (X4.0, Z2.0).
The diametrical XZ coordinates for point A are (X8.0, Z2.0).

EXAMPLE B: Find point B.

1. Start at (X0, Z0).
2. Travel along the Z axis to a point below point B.
3. Move up to point B.

FIGURE 2.20
Absolute coordinate example. The start point is (X0, Z0).

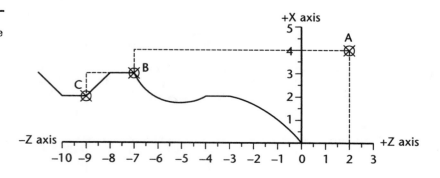

The diametrical XZ coordinates for point B are (X6.0, Z–7.0).
The radial XZ coordinates for point B are (X3.0, Z–7.0)

EXAMPLE C: Find point C.

1. Start at (X0, Z0).
2. Travel along the Z axis until you are below point C.
3. Move up the X axis until you are at point C.

The diametrical XZ coordinates for point C are (X4.0, Z–9.0).
The radial XZ coordinates for point C are (X2.0, Z–9.0).

INCREMENTAL COORDINATES FOR TURNING

The second method for finding points in a Cartesian coordinate system is by using incremental coordinates. Though this method is rarely used for main program coding, it is used with both canned cycles and subroutines.

Incremental coordinates use each successive point to measure the next coordinate. Instead of constant references back to the origin, the incremental method refers to the previous point, like stepping stones across a lake. The following section explains how to find incremental coordinates, again for both the radius and the diameter of the workpiece. (see Fig. 2.21).

REMEMBER

FINDING INCREMENTAL COORDINATES

Starting with the origin, each point in turn is the reference point for the next coordinate. This method is easier to use when you are plotting many closely placed points.

Keep in mind that some controllers use G90 and G91 to change the controller mode between absolute and incremental, whereas other, older controllers do not have G90 and G91. Instead they use X and Z for absolute programming and more commonly U and W for incremental programming. To accommodate both standards, this controller will accept both methods of programming (G90 absolute is the default).

EXAMPLE A: Find point A.

1. Starting at the origin, travel along the Z axis until you are below point A.
2. Move up the X axis until you reach point A.

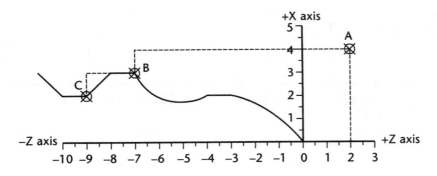

FIGURE 2.21
Incremental coordinate example. The start point is (X0, Y0).

The diametrical XZ coordinates for point A are (X8.0, Z2.0).
The radial XZ coordinates for point A are (X4.0, Z2.0).

EXAMPLE B: Find point B.

1. Starting at point A, travel along the X axis until you are below (or above) point B.
2. Move up (or down) the X axis until you are at point B.

The diametrical XZ coordinates for point B are (X–2.0, Z–9.0).
The radial XZ coordinates for point B are (X–1.0, Z–9.0).

EXAMPLE C: Find Point C.

1. Starting at point B, travel along the Z axis until you are below (or above) point C.
2. Move up (or down) the X axis to find the X coordinate.

The diametrical XZ coordinates for point C are (X–2, Z–2)
The radial XZ coordinates for point C are (X–1, Z–2)

EXERCISES

Refer to Fig. 2.22 to complete the following exercises.

EXERCISE 1: Using Incremental Coordinates
Find the diametrical X and Z coordinates for points A through E.

A: X_____, Z_____   B: X_____, Z_____

C: X_____, Z_____   D: X_____, Z_____

E: X_____, Z_____

EXERCISE 2: Using Absolute Coordinates
Find the X and Z coordinates for points A through E.

A: X_____, Z_____   B: X_____, Z_____

C: X_____, Z_____   D: X_____, Z_____

E: X_____, Z_____

FIGURE 2.22

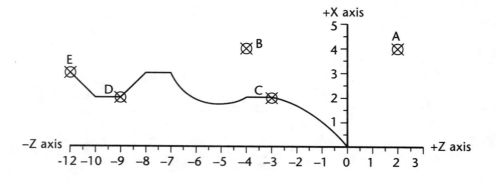

LAB EXERCISES

1. What is the standard coordinate system called?

2. What are the three axes used on the CNC mill?

3. What are the two axes used on the CNC lathe?

4. What are the two types of coordinate systems? Explain the differences between them.

5. On a vertical CNC milling machine, does the X axis run vertically or horizontally?

6. What are the three planes in the Cartesian coordinate system?

7. What is the PRZ?

8. Where do you find the PRZ on a

   milling workpiece?

   turning workpiece?

CHAPTER 3

Programming Concepts

CHAPTER OBJECTIVES

After studying this chapter, the student should have knowledge of the following:

Format of a CNC program

How to prepare to write CNC programs

Types of tool motion

Uses of canned cycles

Tooling

INTRODUCTION

Before you can fully understand CNC, you must first understand how a manufacturing company processes a job that will be produced on a CNC machine. The following is an example of how a company may break down the CNC process.

IMPORTANT

FLOW OF CNC PROCESSING

1. Obtain or develop the part drawing.
2. Decide what machine will produce the part.
3. Decide on the machining sequence.
4. Choose the tooling required.
5. Do the required math calculations for the program coordinates.
6. Calculate the speeds and feeds required for the tooling and part material.
7. Write the NC program.
8. Prepare setup sheets and tool lists.
9. Send the program to the machine.
10. Verify the program.
11. Run the program if no changes are required.

PREPARING A PROGRAM

A program is a sequential list of machining instructions for the CNC machine to execute. These instructions are CNC code that contains all the information required to machine a part, as specified by the programmer.

CNC code consists of blocks (also called lines), each of which contains an individual command for a movement or specific action. As with conventional machines, one movement is made before the next one. This is why CNC codes are listed sequentially in numbered blocks.

The following is a sample CNC milling program. Note how each block is numbered and usually contains only one specific command. Note also that the blocks are numbered in increments of 5 (this is the software default on startup). Each block contains specific information for the machine to execute in sequence.

Workpiece Size: X4, Y3, Z1

Tool: Tool #3, 3/8" Slot Drill

Tool Start position: X0, Y0, Z1.0

```
%                        (Program Start Flag)
:1002                    (Program #1002)
N5  G90 G20 G40 G17      (Block #5, Absolute in Inches)
N10 M06 T3               (Tool Change to Tool #3)
N15 M03 S1250            (Spindle on CW at 1250 RPM)
N20 G00 X1.0 Y1.0        (Rapid over to X1.0, Y1.0)
N25 Z0.1                 (Rapid down to Z0.1)
N30 G01 Z-0.125 F5       (Feed down to Z-0.125 at 5ipm)
```

```
N35  X3.0 Y2.0 F10.0          (Feed diagonally to X3.0, Y2.0
                               at 10ipm)
N40  G00 Z1.0                 (Rapid up to Z1.0)
N45  X0 Y0                    (Rapid over to X0, Y0)
N50  M05                      (Spindle Off)
N55  M30                      (Program End)
```

CNC CODES

There are two major types of CNC codes, or letter addresses, in any program. The major CNC codes are called G-codes and M-codes.

IMPORTANT

G-codes are preparatory functions, which involve actual tool moves (for example, control of the machine). These include rapid moves, feed moves, radial feed moves, dwells, roughing, and profiling cycles.

M-codes are miscellaneous functions, which include actions necessary for machining but not those that are actual tool movements (for example, auxiliary functions). These include actions such as spindle on and off, tool changes, coolant on and off, program stops, and related functions.

Other *letter addresses* are variables used in the G- and M-codes to make words. Most G-codes contain a variable, defined by the programmer, for each specific function. Each designation used in CNC programming is called a letter address.

The letters used for programming are as follows:

N   Block Number: Specifies the start of a block
G   Preparatory function, as previously explained
X   X Axis Coordinate
Y   Y Axis Coordinate
Z   Z Axis Coordinate
I   X Axis location of Arc center
J   Y Axis location of Arc center
K   Z Axis location of Arc center
S   Sets the spindle speed
F   Assigns a feedrate
T   Specifies tool to be used
M   Miscellaneous function, as previously explained
U   Incremental coordinate for X axis
V   Incremental coordinate for Y axis
W   Incremental coordinate for Z axis

REMEMBER

The process of writing CNC programs is primarily the same as going through the steps involved with conventional machining. First, you must decide which units will be used—metric or inch—and which coordinate system will be used—absolute or incremental. Next, a tool must be called up and the spindle turned on. Finally, the tool must move rapidly to a point close to the part to start the actual machining.

These steps are identical in both conventional and CNC machining. The two methods differ only in that in CNC machining the steps are programmed into each CNC file.

Without adequate preparation, a beginner is virtually "doomed" from the start. To avoid this fate, remember that before you write your program you must develop a sequence of operations. Do all the necessary math calculations; then choose your tooling, units, and coordinate system.

IMPORTANT

THREE MAJOR PHASES OF A CNC PROGRAM

The following shows the three major phases of a CNC program.

```
%
:1001
N5  G90 G20
N10 M06 T2
N15 M03 S1200
N20 G00 X1.00 Y1.00
N25 Z0.125
N30 G01 Z-0.125 F5.0
N35 G01 X2.0 Y2.0
N40 G00 Z1.0
N45 X0 Y0
N50 M05
N55 M30
```

1: PROGRAM SETUP

The program setup contains all the instructions that prepare the machine for operation.

%	Program start flag
:1001	Four-digit program number
N5 G90 G20	Use absolute units and inch programming
N10 M06 T2	Stop for tool change, use Tool #2
N15 M03 S1200	Turn the spindle on CW to 1200 rpm

The program setup phase is virtually identical in every program. It always begins with the program start flag (% sign). Line 2 always has a program number (up to four digits, 0–9999; some controllers, however, may have five or six digits). The program number must be preceded by a ":" or an "O" (the letter O).

Line 3 is the first that is actually numbered. Note how it begins with N5 (N for number, 5 for block number 5). You can use any numeric sequence incrementing upward. Throughout this manual we use increments of 5 in our examples. Incrementing in this way enables you to insert up to four new lines between lines when you are editing the program. Some programmers use increments of 1 or 10. The software included with this workbook allows automatic numbering in increments specified by the user.

Block 5 tells the controller that all distances (X, Y, and Z coordinates) are absolute, that is, measured from the origin. It also tells the controller that all coordinates are measured in inch units.

The setup phase may also include commands such as coolant on, cutter compensation cancel, or stop for tool change. Note that different machine tool manufacturers may have specific codes required for specific program setups.

REMEMBER

The three phases of a CNC program are:
(1) Program setup
(2) Material removal
(3) System shutdown

2: MATERIAL REMOVAL

The material removal phase deals exclusively with the actual cutting feed moves.

N20 G00 X1.0 Y1.0	Rapid to (X1, Y1) from origin
N25 Z0.1	Rapid down to Z.1 inches just above part
N30 G01 Z-0.125 F5.0	Feed down to Z–0.125 inches at 5 ipm
N35 X2.0 Y2.0	Feed diagonally to X2 and Y2
N40 G00 Z1.0	Rapid up to Z1 (clear the part)
N45 X0 Y0	Rapid back home to X0, Y0

It contains all the commands that designate linear or circular feed moves, rapid moves, canned cycles such as grooving or profiling, or any other function required for that particular part.

3: SYSTEM SHUTDOWN

The system shutdown phase contains the G- and M-codes that turn off all the options that were turned on in the setup phase.

N40 M05	Turn the spindle off
N45 M30	End of program

Functions such as coolant and spindle rotation must be shut off prior to removal of the part from the machine. The shutdown phase also is virtually identical in every program.

EXAMPLE: Examine the following program and observe how it is written.

This program also shows that some G-codes are **modal.** *Modal* means that a code remains active once executed and remains active until overridden by a different G-code. See blocks N55 and N70.

Note also that leading and trailing zeros are optional and do not affect the outcome of programs when you are using the CNCez simulators—they affect only the presentation. (Refer to blocks N20–N30.) On most CNC controllers, though, the omission of decimal places signifies the smallest machine unit. For example, X1 would actually mean X 0.0001 of an inch if the code is in inch mode. Therefore it is good practice to include decimal points when programming.

%	Program Start Flag
:1001	Program Number
N5 G90 G20	Use Absolute Coordinates and inch programming
N10 M06 T1	Tool change, use Tool #1.
N15 M03 S1200	Turn spindle on CW at 1200 RPM
N20 G00 X1,0 Y1.0 Z.125	Rapid move to X1, Y1, Z.125
N25 G01 Z-0.125 F5.0	Feed down into the part 0.125" at 5 ipm
N30 G01 X3.0	Feed to X3 (still at 5 ipm)
N35 G01 Y2.0	Feed to Y2 (still at 5 ipm)
N40 G01 X1.0	Feed back to X1
N45 G01 Y1.0	Feed back to Y1
N50 G01 Z-.25	Feed down to Z–.25" (still at 5 ipm)
N55 G01 X3	Feed across to X3
N60 Y2	Feed to Y2 (the G01 is MODAL)
N65 X1	Feed back to X1 (G01 is still MODAL)

REMEMBER

Use decimal places to ensure accurate units. Most CNC machines require decimal places.

```
N70  Y1.0              Feed to start point at Y1
N75  G00  Z1.0         Rapid to Z1 clearance
N80  X0  Y0            Rapid tool to home position
N85  M05               Turn spindle off
N90  M30               End of program
```

USING A PROGRAMMING SHEET

You use the CNC program sheet to prepare the CNC program. Doing so simplifies the writing of the CNC program. Look at the sample programming sheet in Fig. 3.1 to see how it works. Each row contains all the data required to write one CNC block. Several blank sheets for programming use are included in Appendix C.

FIGURE 3.1
Sample programming sheet.

NC PROGRAMMING SHEET		PART NAME:				PROG BY:				
		MACHINE:				DATE:		PAGE:		
		SETUP INFORMATION:								
N SEQ	G Code	X Pos'n	Y Pos'n	Z Pos'n	I J K Pos'n	F Feed	R Radius or Retract	S Speed	T Tool	M Misc
5	20, 90									
10									2	6
15								1200		3
20	0	0	0							
25				0.1						
30	1			−0.1		2				
35	1	1.5								

BLOCK FORMAT

IMPORTANT

Block format is often more important than program format. It is vital that each block of CNC code be entered into the CPU correctly. Each block comprises different components, which can produce tool moves on the machine.

Following is a sample block of CNC code. Examine it closely and note how it is written.

N135 G01 X1.0 Y1.0 Z0.125 F5.0

N135	Block Number	Shows the current CNC block number.
G01	G-Code	The G-code is the command that tells the machine what it is to do—in this case, a linear feed move.

X1.0 Y1.0 Z0.125	Coordinate	This gives the machine an endpoint for its move. X designates an X axis coordinate. Y designates a Y coordinate. Z designates a Z coordinate.
F5.0	Special Function	Any special function or related parameter is to be included here. In this case, a feed rate of 5 inches per minute is programmed.

There are some simple restrictions to CNC blocks:

IMPORTANT

- Each may contain only one tool move.
- Each may contain any number of nontool move G-codes, provided they do not conflict with each other (for example, G42 and G43).
- Each may contain only one feedrate per block.
- Each may contain only one specified tool or spindle speed.
- The block numbers should be sequential.
- Both the program start flag and the program number must be independent of all other commands.
- The data within a block should follow the sequence shown in the above sample block, N-block number, G-code, any coordinates, and other required functions.
- Each may contain only one M-code per block.

PREPARING TO PROGRAM

Before you write a CNC program, you must first prepare to write it. The success of a CNC program is directly related to the preparation that you do before you write the CNC program.

REMEMBER

    You should do three things before you begin to write a program:

1. **Develop an order of operations.**
2. **Do all the necessary math and complete a coordinate sheet.**
3. **Choose your tooling and calculate speeds and feedrates.**

DEVELOP AN ORDER OF OPERATIONS

Before you begin writing your program, you should plan it from start to finish, considering all the operations that must be performed. Doing so will help you as you write your program.

DO ALL THE NECESSARY MATH AND COMPLETE A COORDINATE SHEET

Do all the math that has to be done before you begin the program. You can mark up your part drawing and use a coordinate sheet, as illustrated in Fig.3.2, if you find it helpful.

FIGURE 3.2
Sample coordinate sheet.

#	X	Z
1		
2		
3		
4		
5		
6		
7		
8		
9		
10		
11		
12		
13		
14		

CHOOSE YOUR TOOLING AND CALCULATE SPEEDS AND FEEDRATES

Decide on which tools you are going to use and verify that the tools you have available will perform the required tasks. Also, calculate the required speeds and feedrates that you will be using in your program.

FIGURE 3.3
Sample setup sheet.

Setup Sheet				
Cutting Tools				
Station #	Tool Description	Catalog Number	Insert	Comments
1				
2				
3				
4				
5				
6				
7				
8				
9				
10				
11				
12				

Setup Instructions

Setup Sketch

You can use a setup sheet, such as the one shown in Fig. 3.3, to help you choose your tools and to help you with the setup required before actual machining.

You can also use a coordinate and setup sheets to help prepare jigs and fixtures. These sheets are helpful, as well, when you are modifiying registers and parameters in the machine's controller and preparing the machine's tool magazine for a particular job.

PROGRAM ZERO

Program zero allows you to specify a position from which to start or to work. Once program zero has been defined, all coordinates that go into a program will be referenced from it. When you work from a constant program zero, you are using absolute programming, as discussed in Chapter 2. In incremental programming, you have in effect a floating program zero that changes at all times, whereby the current position acts as the reference for the next move. However, for most of the program samples in this workbook, absolute programming is used with a fixed program zero, or origin. To specify absolute positions in the X direction we use the X-address word. To specify absolute positions in the Y and Z directions we use the Y- and Z-address words, respectively. The position that we select for milling is always the lower left-hand corner and top surface of the workpiece (see Fig. 3.4). The position we use for the lathe is always the center of the part in X and the right-hand end of the finished workpiece in Z (see Fig. 3.5).

REMEMBER

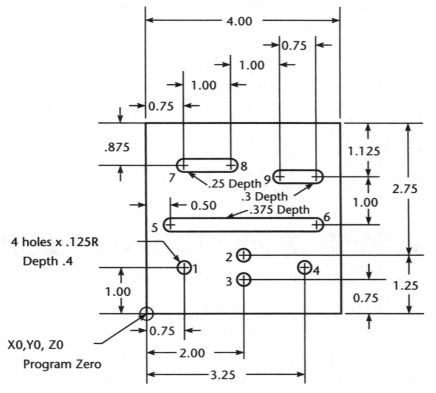

FIGURE 3.4
Milling program zero location.

#	X	Y	Z
1			
2			
3			
4			
5			
6			
7			
8			
9			
10			

Coordinate Sheet

FIGURE 3.5
Lathe program zero location
X coordinates are in diameter
measurements.

Coordinate Sheet

#	X	Z
1		
2		
3		
4		
5		
6		
7		
8		
9		
10		
11		
12		
13		
14		

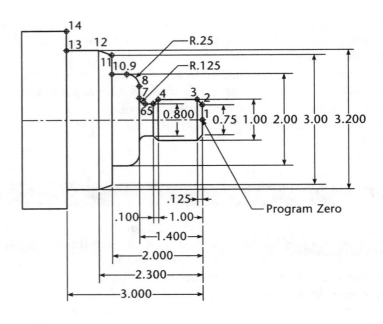

TOOL MOTION

IMPORTANT

Generally, three types of tool motion are used on a CNC machine.

G00 Rapid tool move. Nonmachining command. Each axis trajectory is exhausted as fast as the motor can drive the axes.

G01 Straight-line feed move. Linear interpolation. Coordinated moves at a controlled feedrate.

G02/G03 Two-dimenional arc feed moves. Circular interpolation.

Other tool motions include helical and multiplane 3D arc moves.

All cycles, such as G71 rough turning or G81 drilling, are either one of these types or a combination of these types of motion.

These motion commands are *modal*. That is, if you program one of these commands, you do not need to program the same code again until you want to change the type of tool motion.

Other codes in CNC programming also are modal and will be identified as you learn the various commands. These other codes are explained in detail in the following chapters.

USING CANNED CYCLES

Canned, or fixed program, cycles are aids that simplify programming. Canned cycles combine many standard programming operations and are designed to shorten the program length, minimize math calculations, and optimize cutting conditions to improve the efficiency of the machine.

Examples of canned cycles on a mill are drilling, boring, spot facing, tapping, and so on; on a lathe, threading, rough facing and turning, and pattern repeating cycles. On the lathe, canned cycles are also referred to as multiple repetitive cycles. You will find examples of these cycles as you work through the milling and turning sections of this book. Examples that you can refer to quickly in each section are the G81 drill cycle in milling and the G71 rough turn cycle in turning that are presented in Chapters 5 and 6.

All these cycles speed up programming. Also, you should always be familiar with the canned cycles that your CNC control offers.

Subroutines are also available on many CNC controllers. Though these are not canned cycles, you can use them to code your own specialized routines, which can be considered personal canned cycles.

TOOLING

Not all cutting operations can be performed with a single tool. Separate tools are used for roughing and finishing, and tasks such as drilling, slotting, and thread cutting require their own specific tools.

The correct cutting tool must be used at all times. The size and shape of the cutting tools that you can use depend on the size and shape of the finished part. A tool manufacturer's catalog will give you a complete list of the various types of tools available and the applications for which each should be used. Remember that the depth of cut that can be taken depends on the workpiece material, the coolant, the type of tool, and the machine tool itself. The following must be taken into consideration when choosing your tools.

Modern drills and end mills with carbide inserts. (Courtesy of Carboloy, Inc., USA.)

DRILLING

The tool most commonly used to make holes is the fluted drill. Drills are made with two, three, or four cutting lips. The two-lip drill is used for drilling solid stock. The three- and four-lip drills are used for enlarging holes that have been previously drilled.

MILLING

On a lathe, the cutting tool is fixed and the work rotates. In a milling machine, the cutter rotates and the work is fed against it. The rotating cutter, termed the milling cutter, has almost an unlimited variety of shapes and sizes for milling regular and irregular forms. The most common milling cutter is the end mill. Other tools that are often used are shell mills, face mills, and roughing mills. When milling, you must take care not to make a cut that is deeper than the milling cutter can handle.

End mills come in various shapes and sizes, each designed to perform a specific task. The three basic shapes of standard end mills are flat, ball-nose, and bullnose.

PUNCHING AND NIBBLING

Dies are used in CNC punching and nibbling machines. Repeated punches along a path produce a nibbling effect in order to cut complex shapes in sheets of metal. The metal gauges are sometimes very heavy and would otherwise be impossible to cut with conventional hand cutting tools. Modern CNC punch machines are capable of extremely fast punch rates. Most have indexable punch turrets with several die shapes per punch head.

Several configurations of end mills in varying sizes for different applications. (Courtesy of Carboloy, Inc., USA.)

Various indexable punch tools. (Courtesy Strippit, Inc.)

TURNING

In lathe operations, the tool is driven through the material to remove chips from the workpiece in order to leave geometrically true surfaces. The type of surface produced by the cutting operation depends on the shape of the tool and the path it follows through the material. When the cutting edge of the tool breaks down, the surface finish becomes poor and the cutting forces rise. Vibration and chatter are definite signs of tool wear, although many factors such as depth of cut, properties of materials, friction forces, and rubbing of the tool nose also affect tool vibration.

Modern turning tools with carbide inserts. (Courtesy of Carboloy, Inc., USA.)

FIGURE 3.6
Several common turning tool designations in the ANSI B212.3–B212.5 standards.

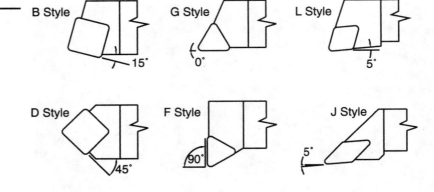

The following factors determine how a cutting tool performs:

The tool material
The shape of the tool point
The form of the tool

No one cutting material is best for all purposes. The principal materials used in NC tooling are carbon tool steel, high speed steel, cast nonferrous alloys, carbides, sintered oxides, diamonds, artificial abrasives, and coated tooling.

Lathe tool shape and form should also be considered when you are choosing your tooling. The tool angles, relief angles, rake angles, cutting edge angles, and tool nose radius all affect the way that metals are cut.

Most modern CNC turning centers utilize an indexable tool changer that holds turning tools with carbide inserts. The use of indexable insert holders for these lathes adhere to the American National Standards Institute (ANSI) standards, B212.3, B212.4, and B212.5 (see Fig. 3.6). These standards specify insert radius compensation, holders, and insert styles. Carbide inserts are disposable and come in various sizes and strengths. Ten positions are used in the insert style identification system, ANSI B212.4. Each position specifies particular characteristics of the insert. Positions 1 through 7 include shape, relief angle or clearances, tolerances, type, size, thickness, and cutting point configuration, respectively. These are all standard identification items. Positions 8 through 10 are used only when needed.

For example, an insert with a diamond shape, 30Þ relief angle, a tolerance of 0.005–0.010 inch, chip grooves on both surfaces, and a hole, the size of a 1/2 inch inscribed circle, and a 1/2 inch thick, 1/32 inch tool-nose radius would be designated as

$$
\begin{array}{ccccccc}
1 & 2 & 3 & 4 & 5 & 6 & 7 \\
D & G & M & F & 4 & 8 & 2
\end{array}
$$

FEEDRATES AND SPINDLE SPEEDS

Every material differs in characteristics: from softer materials such as machineable wax to harder materials such as stainless steel. New tool technology has produced a wide range of tools that can be used at greater speeds and feeds for longer periods. It is very important that you fully understand the value of the correct spindle speed and feedrate. Too fast a speed or feedrate will result in early tool failure or poor surface finish. Too slow a speed or feedrate will lead to increased machining time and, possibly, greater part cost. A good surface finish and economical production rates require the proper use of spindle speeds and feedrates. The spindle speed and feedrates are influenced by several conditions, as well as by the power and stability of the particular machine tool.

The spindle speed is the peripheral speed of the work passing the cutter (turning) or the rotating cutter passing the surface of the part (milling). Appendix B contains detailed information on cutting speeds and feedrates.

For milling, the correct speeds and feedrates are determined in part by the diameter of the cutter, spindle RPM, number of teeth on the cutter, chip load per tooth, and surface feet per minute for a particular material.

Refer to the *Machinist's Handbook* to find the specific feedrates and speeds for a particular material and cutter. Appendix D shows a sample table from the *Machinist's Handbook*.

CUTTING FLUIDS

There are three main reasons for using cutting fluids:

To remove or reduce the heat being produced

To reduce cutting tool wear

To help clear chips from the workpiece area

Thus in most applications fluids are used for cooling and lubrication. However, in some cases they also are used to help in the removal of chips, to prevent the production of dust or irritants, and to dampen machine vibration. Other lubrication may also be applied for specific operations, such as the use of tapping compounds during tapping operations.

Ordinarily, a general purpose cutting fluid solution for milling, turning, and drilling ferrous metals consists of 1 part soluble oil to 20 to 30 parts water. For tool steels and hard alloys, a heavier 1 part oil to 10 parts water is recommended. For drilling, reaming, or tapping of extremely hard high-strength steels, a sulferized or sulfochlorinated mineral oil is recommended. New advances in cutting-fluid research have made available new environmentally friendly lubricants.

Always use the best cutting fluid for the current job.

REMEMBER

LAB EXERCISES

1. What do the following letter addresses represent?

   X:
   Y:
   Z:
   F:

2. What are the basic definitions of the letter addresses?

   G:
   M:

3. What is a preparatory function?

4. What are the two reasons for using cutting fluids?

5. What are the factors that affect how a cutting tool performs?

6. Describe the typical PRZ for milling.

CHAPTER 4

Interactive Simulation Software

After studying this chapter, the student should have knowledge of the following:

- The user interface of the CNC simulation software
- How to install the simulation software
- How to use the interactive CNC editor
- How to run a simple CNC simulation

INSTALLATION AND SETUP

Before attempting to install the software, verify that your workstation has the following:

100 MHZ or higher Pentium class CPU

Windows 98, NT 4.0 or higher

SVGA or better graphics monitor 800 × 600 × 256 res. or higher

120 MB free hard disk space (60MB minimum)

64 MB RAM (32MB minimum)

8X CD-ROM or faster (12X minimum)

INSTALLING AND RUNNING THE WORKSHOP

1. Insert the CD in your CD-ROM drive. If Autostart is enabled for your CD-ROM drive, then the setup program will automatically start. If not, from the task bar select Start and then Run. In the Open field enter x:\start, where x is the drive letter of your CD-ROM. The CNC Workshop start program window will be displayed, as shown in Fig. 4.1. Select the Start option. The program will automatically check to determine whether all of the necessary components have been installed to launch the CNC Workshop. They may include your currently installed software versions of Internet Explorer, the JAVA Virtual Machine, and other components necessary to view the CD content properly.

2. If the Run option returns a dialog stating that not all components have been installed, select the Install option. The CNC Workshop

FIGURE 4.1
The CNC Workshop splash page. Here you can either run or install the CNC Workshop.

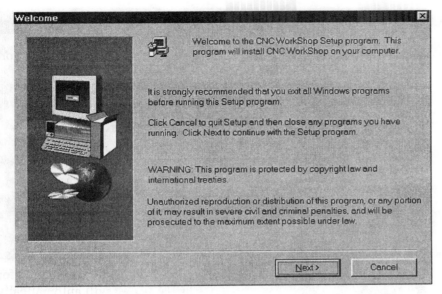

FIGURE 4.2
The CNC Workshop installation
program Welcome Screen.

installation program will then be launched. Follow the installation instructions as they are displayed on the screen (see Fig. 4.2). All the components required to run the CNC Workshop will be installed on your computer's hard disk. The CD-ROM includes Internet Explorer, because the CNC Workshop CBT is browser-based, as well as plug-ins for proper viewing of the CBT content.

3. After the installation has been completed, you may have to reboot your system for all the changes to take effect. To run the CNC Workshop, from the task bar select Start, Programs, Addison-Wesley, CNC Workshop, and CNC Workshop CBT. Alternatively, if you chose to add a shortcut to your desktop, just select the CNC Workshop shortcut icon.

4. When the CNC Workshop program is executed, a splash screen will appear (see Fig 4.3). This program will launch the Internet Explorer and load the CNC Workshop CBT start page (see Fig. 4.4). From this main screen you can run the CNC Workshop CBT course or launch the included CNC simulators. Moving your mouse pointer over the options will highlight them.

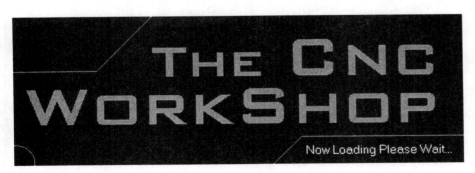

FIGURE 4.3
The CNC Workshop splash screen
is displayed prior to loading the
CBT start page.

FIGURE 4.4
The CNC Workshop CBT start page. Here we launch the CBT content or run the CNCez simulators.

DEFAULT CONFIGURATION

Certain standards and modes are defaults in this program for both milling and turning—for example, diametrical programming for turning. To change many of the defaults for either milling or turning, select Preferences... from

FIGURE 4.5
The CNCez Milling Simulator Machine Configuration Variables dialog screen. Here custom user preferences may be set and saved.

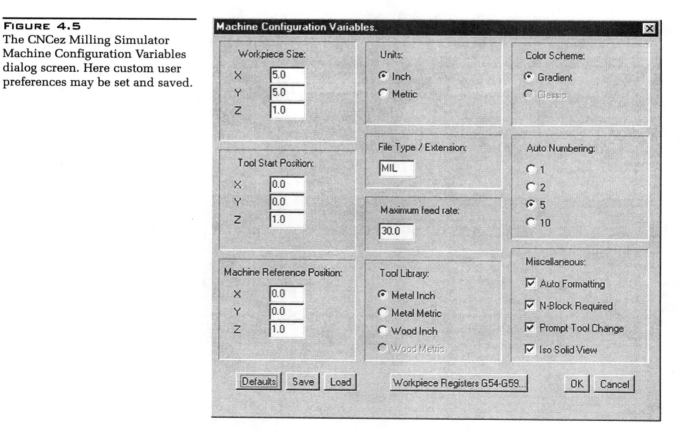

the File menu. A default configuration dialog will be displayed, allowing you to change the default Machine Configuration Variables and save and load your own personalized configuration setups (see Fig. 4.5).

THE USER INTERFACE

The user interface is divided into different areas, each dedicated to a different function (see Fig. 4.6). As the program information is updated, so are the different screen areas.

The simulation environment has been developed to provide the user with the maximum amount of relevant program information without screen clutter and data overload. Each window displays current information about the program in real-time simulation. As the data are updated in a running program, the respective windows display the appropriate updated information.

The following describes the content of each screen area.

Menu Bar
Contains the pull-down menus that govern system operations. Use the cursor or the respective F-key to pull down a menu.

Identification Line
Displays the current file name, as well as current workpiece dimensions and units of measurement.

Tool Display Window
Displays a graphic representation of the tool currently in use.

Machine Status Window
Defaults (factory set) to the following settings. All settings are user-definable. When they are altered, the new settings are displayed.

Units	Inch (can be changed to mm)
Arc Plane	XY (can be changed to XZ or YZ)
Positioning	Machine positioning (Absolute;

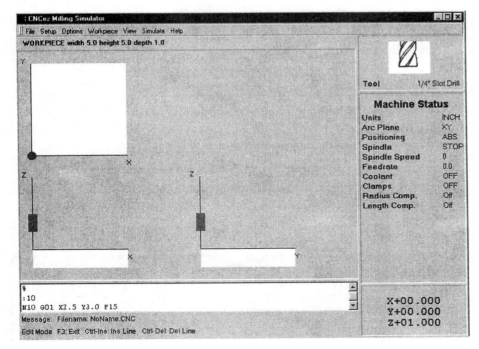

FIGURE 4.6
The CNCez Milling Simulator graphic user interface.

incremental optional)

Spindle	Stop (can be turned On)
Spindle Speed	Indicates the current spindle speed
Feedrate	Indicates the current machine feedrate
Cycle	(Displays current cycle, if any)
Coolant	Off (can be turned On)
Opt. Stop	(Displays optional stop status)
Blk. Skip	(Displays block skip status)
Clamps	Off (can be switched On)
Radius Comp.	Displays the radius compensation status
Length Comp.	Displays the length compensation status

Coordinate Display Window

Displays the current tool position. In default mode the panel shows the absolute position. In the alternative mode a left column shows absolute coordinates; and a right column shows incremental coordinates (see Fig. 4.7). You can access this option by a right-click on the coordinate display window.

FIGURE 4.7
Selecting the alternative coordinate display pop-up menu.

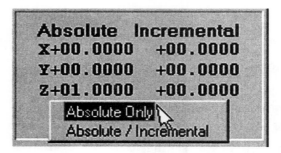

Graphics Simulation Window

Displays the program simulation. As each block is executed, the window displays the toolpath and material removal. Different views are available. (Refer to the View menu.)

Message Line

Displays the block currently being executed and a simple English translation of what the code means.

ACCESSING MENU OPTIONS

The entire program is controlled by the pull-down menus and the options within them, so you need to be able to access them. For this version of the simulators, the mouse cursor is used to access menus.

1. Simply use the mouse. Move the mouse around until the mouse cursor (arrow) overlaps the title of the desired menu. Clicking on the Menu Bar item (see Fig. 4.8) will open the pull-down menu.

FIGURE 4.8
The CNCez Simulator's Menu Bar

CNCez Milling Simulator

File Setup Options Workpiece View Simulate Help

2. To select an option from within a pull-down menu, move the cursor until it highlights the selected option. When the option is highlighted, press the left mouse button.

MENUS

FILE

The familiar Windows style File Menu used in the CNCez simulators is shown in Fig. 4.9.

FIGURE 4.9

New
Allows you to start a new part program. It clears the current part program from the memory buffers and resets the screen environment for a new part program. If the old part program has not yet been saved, a screen prompt appears asking if you want to save it.

Open...
Opens a familiar Windows File Open dialog as shown in Fig. 4.10. You can retrieve an existing program from a folder on your hard drive or floppy disk. Once loaded the file may be used for editing or simulation.

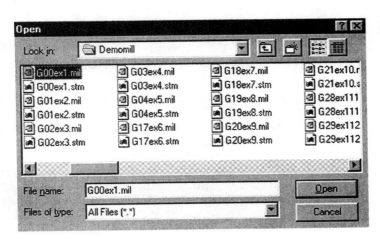

FIGURE 4.10

Note, all example programs are located in the Demomill or Demoturn folders.

Save

Saves the CNC code for the current part program, as well as the environment data. (Environment data comprise any information about graphics, part dimensions, tool information, and so on.) These are saved as different files. The CNC code is saved as an ASCII text file with the .mil or .trn extension as a default extension unless changed in the defaults configuration dialog.

Save As...

Saves the current CNC file under a new name. After you select this option, the familiar Windows Save As dialog will appear wherein you can enter the filename or select a file to save to. If you enter a name that already exists, you are prompted "File exists—Overwrite?" whereupon you select either Yes or No.

Edit

Puts you into the CNC text editor in which you can input and edit CNC codes. F1 toggles between the full screen and a three-line screen that displays much more of the current CNC file.

When in the Edit mode the following control keys are used to edit the CNC program.

Cursor Movements

Left/Right Cursor Key	Moves the cursor left/right.
Up/Down Cursor Key	Moves the cursor up/down.
Ctrl Left/Right Arrow	Moves the cursor one word left/right.
Home/End	Moves the cursor to the start/end of the current line.
Shift + Arrow Keys	Used to highlight words or lines.

Delete Functions

DEL	Deletes the character to the right.
BACKSPACE	Erases the character to the left.
Ctrl-DEL	Deletes the entire line.

Miscellaneous

Ctrl-Insert	Inserts a blank line after the current line.
F3	Exits the editor to the current line.

Print

Prints the currently opened program file to the Windows default printer. Not to be confused with the print option in the View, Solid Dialog, this command will print the CNC program as text information only.

Preferences...

Displays the default configuration dialog in which you can specify the preferred default configuration variable settings. (See Fig. 4.5.)

Workpiece Size

The values for X, Y, and Z set the default workpiece size, which will be displayed when you select the New... menu item from the Workpiece menu.

Tool Home Position

This is a default start point for the CNC simulator. You use it to reset the machine's position when restarting a simulation.

Machine Reference Position

This is the reference point that you use when invoking a G28 or G29 command in the mill simulator. Think of it as the tool change position.

Units

You can easily change the default Units mode by selecting either inch or metric units.

File Type/Extension

You can set the default filetype or extension here. For compatibility with the previous CNCez simulator files, the .mil and .trn extensions have been preset. You may want to change these to the more common .cnc or .nc extensions used in many CAM software packages.

Maximum Feedrate

You can use this field to set the maximum feedrate. This is for checking purposes so that no extraneous values are used during program development. It also limits the maximum feedrate used in the controller for safety purposes.

Tool Library

The default tool libraries are set to metal inch. There are four different libraries to choose from. You can create custom libraries simply by editing the appropriate text file. However, this method is reserved for advanced users only.

Color Scheme

The default color scheme for the depth indication is the Gradient method. Here the depth change is indicated by the varying blue color. The darker the blue, the deeper the tool is. The classic scheme will display depths similar to the earlier CNCez simulators.

Auto Numbering

Auto Numbering, when enabled in the Options menu, will use the values set here to auto-increment the CNC block numbers when in the edit modes. The default auto-increment value is set to 5.

Miscellaneous

Auto Formatting, if enabled, will automatically format CNC block lines as they are typed in when in the edit modes.

N-Block Required

N-Block Required, if enabled, will set the simulator to require that N-block numbering be used. If turned off, N-blocks are not required. This is useful when you are simulating programs generated from some CAM packages, especially when large files are produced.

Prompt Tool Change

Prompt Tool Change, when enabled, will bring up a Tool Change dialog prompt when a tool change command is encountered in a CNC program.

Iso Solid View

When the Iso Solid View option is checked, ☑, a real-time mini machined solid is displayed in the graphics area when either the 1st or 3rd View modes are selected from the View menu.

Defaults

The Defaults button will reset all the settings to the system defaults.

Save

The Save button allows you to save the current configuration preferences. The familiar Windows File Save dialog will appear.

Load

The Load button allows you to read a previously saved configuration. The familiar Windows File Open dialog will appear.

Workpiece Registers G54-G59

This button will open another dialog, which holds the G54-G59 Workpiece register settings. They can also be saved and loaded. These are extremely useful in a production environment.

OK

When you select the OK button, all currently displayed configuration variables will be utilized for the current session.

Cancel

The Cancel button will ignore any changes you have made to the configuration variables.

Exit

Exits from the program and returns you to the Windows operating system or the CNC Workshop CBT. If the file has been changed and not saved, you will be asked if you would like to save it first.

SETUP

The Setup menu (see Fig. 4.11) allows you to view and modify the loaded Tool Library and the Offset Register table.

FIGURE 4.11

Mill 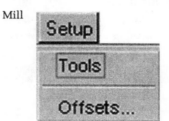   Turn

TOOLS

When you select Tools, a window will be displayed, showing two smaller windows (see Fig. 4.12). The tool library is shown in the larger window on the left and contains 24 standard tools. The tool turret, the smaller window on the right, can contain any of 16 tools from the library in any order, depending on program requirements.

To move a tool from the library into the turret, use the cursor to move

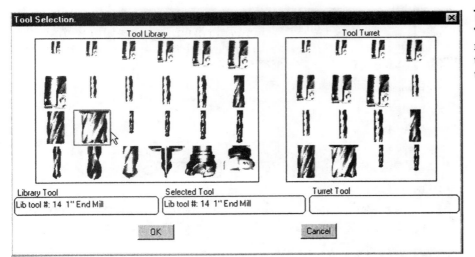

FIGURE 4.12
The Tool Selection dialog for milling. Here you can build a custom turret, using the tools from the current Tool Library.

the green box until it rests on the desired tool. Select this tool by clicking on it; the green box will change to red, indicating that this is now the selected tool. To transfer the tool to the Turret, move to the Turret area and a green box will follow the mouse movement as it did when you were selecting the tool from the library. Move the green box to the desired turret location. Click on the desired location and the green box will again change to red, indicating the change made to the Tool Turret. Once the Tool Turret has been set up, clicking on the OK button will save this Tool Turret for the current session.

When you use different tools in a program, it is the Tool Turret that distinguishes between tools. It may be simpler to use the default Tool Turret, as the tools in it are preorganized.

OFFSET REGISTERS

The Offset Registers dialog boxes display all the current register values as shown in Figs. 4.13 and 4.14. A default set of register values is preset into the simulators. These can be modified to suit a particular job and each set can be saved and recalled at a later time. The register values are used for the tool length and tool radius compensation in the mill and for the tool-nose radius compensation in the lathe CNCez programs. The same registers may be utilized by both the G41/G42 cutter radius compensation and the G43/G44 cutter length offset. Note that the turning offset register dialog is different from that for milling, in that it also includes X-Set and Z-Set values. These are for reference only and are functional only with the controller. Refer to Chapters 5 and 6 for detailed explanations of how the offsets work.

Offset Registers.

Register 0	0.0	Register 6	0.15	Register 12	0.3	Register 18	0.45
Register 1	0.025	Register 7	0.175	Register 13	0.325	Register 19	0.475
Register 2	0.05	Register 8	0.2	Register 14	0.35	Register 20	0.5
Register 3	0.075	Register 9	0.225	Register 15	0.375	Register 21	0.525
Register 4	0.1	Register 10	0.25	Register 16	0.4	Register 22	0.55
Register 5	0.125	Register 11	0.275	Register 17	0.425	Register 23	0.575

Save   Load   OK   Cancel

FIGURE 4.13
The Offset Registers dialog for milling. Here is where the radius and height compesation registers are referenced. Note that Register 0 is always set to 0.0 and cannot be changed. It is usually used as a reset register.

FIGURE 4.14
The Offset Registers dialog for turning. Only the TNR values are used for the simulator.

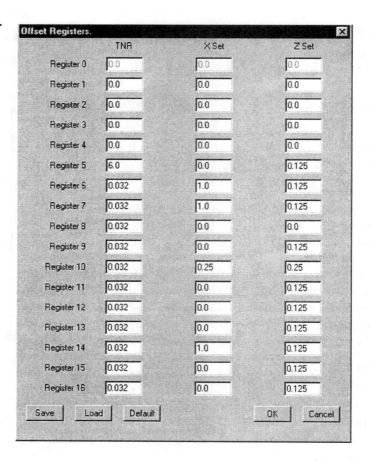

OPTIONS

All items in the Options menu can be toggled on or off simply by selecting that option. A check mark, ☑, indicates that the option is selected, or on, as shown in Fig. 4.15.

Optional Stop

Toggle option that enables or disables the program's optional stop. If Optional Stop is on (☑), the simulation and controller will stop when it encounters an M01 command and remain inactive until Enter is pressed. If Optional Stop is off (☑), the M01 command will be skipped. The default is off (☐) in the screen shot.

Block Skip

Toggle option that enables or disables the block skip lines. A line

FIGURE 4.15

with a Block Skip starts with a "/" (forward slash). If the block skip is on (☑), the lines are not executed. When the block skip is off (☐), these blocks are executed in the program. The default is off as in the screen shot.

Dry Run (milling only)

Toggle option that enables or disables the dry-run mode. If Dry Run is on (☑), all the Z moves are ignored and the spindle will always be off. The default is off (☐) as in the screen shot.

Auto Line Numbering

Toggle option that generates automatic block numbers during the edit mode. When selected, or on, Auto Line Numbering generates automatic block numbers with a block increment as set in the Preferences... option of the File menu. If Auto Line Numbering is off, no block numbers are generated; you must manually enter the block number. (If N-Block Required is set and no block number is entered, the software will generate an error message.) The software allows you to enter a previously used block number. The default is on (☑), with a block increment of 5, as in the screen shot.

JOG MODE [R F S] (controller only)

This options allows you to select the axis speed for movements in Jog. When you select this item, a pop-down menu gives you a choice of speeds required for Jog. The default is Rapid.

Rapid	Highest velocity
Feed	Current set feed
Step	Smallest machine increment

SHOW TOOL PATH

Toggle option that generates a red graphics line that trails the tool center around the simulation screen. It is visible in all three views and provides some useful data regarding the tool's actual position. The default is off (☐), rapid or positioning moves will be dashed lines, and feed or cutting moves will be solid lines.

FULL ANIMATION

This toggle option lets you turn on the Full Animation setting provided the option is on (☑). If Full Animation is off (☐), then simulation results will be rendered on the screen much faster without simulating the complete tool movements.

WORKPIECE

The Workpiece Menu is shown in Fig. 4.16.

New...

Brings up a dialog box wherein you can enter the workpiece dimensions.

For Milling

The workpiece dimensions are X for length, Y for width, and Z for height. Note that the default values are displayed as set in the Preferences... dialog from the File Menu.

FIGURE 4.16

Mill      Turn

For Turning
Prompts you to input stock length, outer diameter, and if required the inner diameter for pipe workpieces.

Load...
Loads a previously saved workpiece. The standard Windows File Open dialog will appear, listing all saved workpiece files available for loading.

Save...
Brings up the familiar Windows Save File dialog, allowing you to save the current workpiece by entering a file name. The workpiece is saved with a different file extension from the program's, so identical file names may be used. Or you may use the dimensions as a file name, for example, 3X1.

Pipe... (turning only)
Allows you to modify the current part blank inner diameter value or to change it to a pipe if no inner diameter was specified in the New dialog.

VIEW

The View menu is shown in Fig. 4.17.

1st Angle (Milling only)
Displays the European standard projection: side view, front view, and plan view.

3rd Angle (Milling only)
Displays the North American standard projection: plan view, side view, and front view.

FIGURE 4.17

Mill   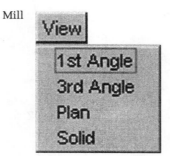   Turn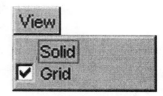

Plan (Milling only)

Displays a plan view of the part only, as illustrated in Fig. 4.18. It is utilized when you are using small cutters or when more clarity is required. Shown at the edge of the simulation window is a depth indicator—a red bar that visually displays cutter depth relative to the part.

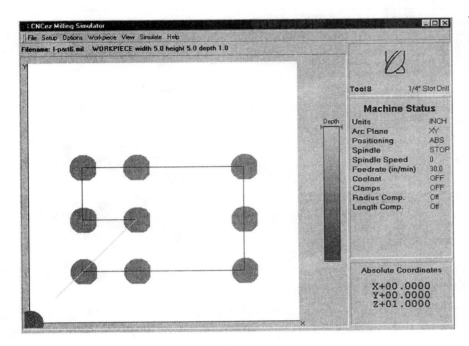

FIGURE 4.18
Note that the deeper the tool is positioned, the darker the shade color becomes.

Solid

Displays a separate window showing a solid three-dimensional view of the part after the program has been executed. The part may be viewed from four different angles: front, left side, right side, and back (see Fig. 4.19). You may also print out this screen by using the Windows default printer.

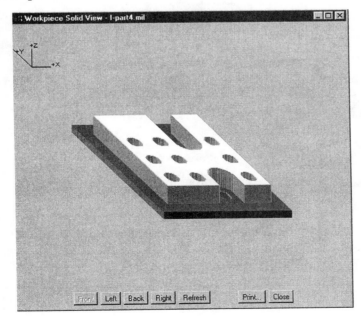

FIGURE 4.19
The Solid Workpiece View window.

Grid (turning only)

Displays as a quick reference a small scale grid on the X and Z axes. This can help you in determining the tool move locations.

SIMULATE

The Simulate menu is shown in Fig. 4.20

FIGURE 4.20

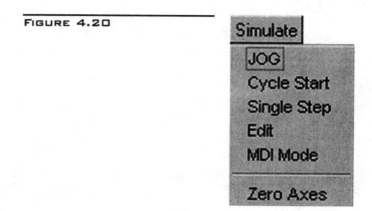

Jog

Simulates tool movement of the machine tool. The cursor keys control tool movement, as well as being used for clicking on the appropriate panel buttons (see Fig. 4.21).

Jog feedrate, spindle speed, and tool selection can be changed by selecting the Tool, Feedrate, or Spindle Speed buttons, or by using the increment and decrement buttons of the panel. The changes will appear on the right of the simulator status panel.

FIGURE 4.21

The Jog panel for milling simulates the Jog functions found on the front panel of most CNC machines.

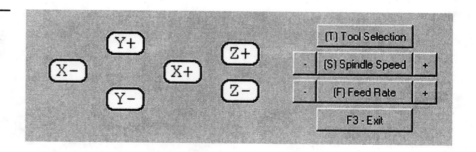

Cycle Start

Executes the entire program. If there are any errors in the program, the system will stop and display the appropriate error message for each. Keep in mind that the simulation speed of the tool is not the actual tool feedrate. This can be set to Full Simulation in the Options menu, which will slow the drawing of each move considerably and is used only for instructional purposes. Note that, if the Prompt Tool Change is enabled in the Preferences... dialog, you will be prompted prior to each Tool Change. Pressing F3 halts the simulation.

Single Step

Simulates the current CNC program line by line. Each line to be executed is highlighted; to continue, press ENTER. Pressing F3 halts the simulation.

Edit

Where all real-time CNC editing occurs. As each line is written in the edit box, it is executed. The resulting actions are animated in the graphics simulation window. The tool status window and program status window keep track of system functions concurrently, constantly being updated. Using Edit, you can edit your CNC program while you are actually executing it. This is a very useful learning tool.

The system also checks for syntax while confirming all tool movements for feasibility. A display prompt informs you of any errors or impossible functions. To correct an incorrect block of CNC code, use the cursor keys to remove the bad line and then reenter it with the correct information.

When editing the program, use the cursor keys to move up, down, left, and/or right. You can move around this way without triggering screen refresh. If you move around and also make a change, moving down a line will cause a refresh.

Edit mode also numbers the CNC blocks automatically in a standard 5-block increment (you may modify this increment amount with the Preferences... dialog) and saves the program changes to memory. When you are satisfied with the program, you may save it for later retrieval via the File Open... dialog.

MDI Mode

Simulates the manual data input (MDI) function. Every block is treated and executed independently, and not as part of a CNC program. Use MDI to move the tool position or test individual blocks. MDI does not automatically number the lines because they are not saved in memory. MDI commands are *one-time executable* and are discarded from memory after execution.

ZERO AXIS

This option is used instead of the G92 command. Use the simulation Jog panel to move the tool to the new origin and then use Zero Axis. A screen prompt will be displayed: "CORRECT TOOL POSITION? Y/N." If you press Y for yes, the system will reset the absolute coordinates to X0.0, Y0.0, Z0.0, the new origin.

Note: Most programs require the tool to be in the tool home position (X0, Y0, Z1) prior to execution. When using Zero Axis, be sure to raise the tool prior to program execution.

ESC (Escape) or F3

Used during program simulation to stop the cycle. When using the Cycle, Step, or Edit mode, pressing ESC or F3 will cause the simulation to stop after the current line has finished executing. A display prompt will ask: "CONTINUE? YES/NO."

HELP

The Help menu items are shown in Fig. 4.22

FIGURE 4.22

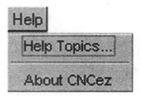

Help Topics...

Lists help topics. Each topic has nested subtopics. Picking an item displays a box of help text or a further submenu.

About CNCez

Displays the program's version number and a copyright notice, as shown in Fig. 4.23.

FIGURE 4.23

GETTING STARTED

This section deals with loading, editing, writing, and executing CNC programs, using the simulation software. There are several sample procedures to follow, which are intended to be a beginner's guide to program manipulation. The more experienced user will explore the endless possibilities of the simulator and gain knowledge en route.

LOADING AND RUNNING A SAMPLE CNC FILE

1. If you have installed the software on your hard disk, from the CNC Workshop CBT main screen select either the Milling or Turning Simulator. Alternatively, you can run the CNCez simulators from the Windows desktop. From the task bar select Start, Programs, Addison-Wesley, CNCez, and either CNCez Milling Simulator or CNCez Turning Simulator.
2. The main simulation screen will now be displayed. You will be prompted to enter the blank part size. The system default sizes will be displayed (the values set in the Preferences... dialog).
3. Access the File menu by moving the mouse cursor over the menu bar item and clicking on it; then scroll down to highlight and click on the Open... option.
4. The familiar Window File Open dialog is then displayed. From the Demomill folder a list of CNC programs is displayed. The Look In:

field should display an open folder icon with either a Demomill or Demoturn name. These folders include all the sample files, including subfolders of various items and machining icons. Select the CNC file that you want to load; then click on the OK button. The CNC file and, if available, any associated workpiece file, will be loaded into memory. The workpiece dimensions and the current file name will be displayed in the information bar below the menu bar.

5. To start the simulation, access the Simulate menu and then highlight and select the Cycle Start option. The program is simulated on screen. (Be sure to watch the screen for any user prompts, such as for a tool change.)

This entire procedure can be condensed into the following commands:

File, Open...
Pick File...
Simulate, Cycle Start

EDITING AN EXISTING PROGRAM

1. Once the CNC file has been opened and is residing in the simulation environment's memory, highlight and select Simulate Edit.
2. The CNC file is now displayed in the text editor box. Use the cursor keys to scroll up or down or press F1 to toggle between the Full Screen and three-line editor options.
3. To exit the Full Screen text editor, press F3, as displayed at the bottom of the screen.

Note: It is important to realize that any changes to the program made in the text editor will be checked for syntax and errors to ensure correct CNC code. The program will simulate each line only in the Simulate menu.

WRITING AND EXECUTING YOUR OWN CNC PROGRAM

1. Access the File menu and select New....
2. In the File Open dialog, type in the alphanumeric file name of the program that you are about to create and write; then press Enter or click on OK.
3. To choose the Units option, access File, Preferences.... In the Machine Configuration dialog, select the desired units mode in the Units group box. Also, select any other preferences you want to work with here, such as default workpiece size and the tool home position of the machine reference.
4. To define a workpiece, access the Workpiece menu and select the New... option. A workpiece definition dialog will be displayed. Here you can accept or edit the appropriate fields for the X, Y, and Z dimensions. These correspond to the width, length, and height of the workpiece, respectively. When the new dimensions have

been entered, click on the OK button and a new workpiece will be defined in the same units defined in the Machine Configuration default menu.

5. To change the view, access the View menu and select a view; 3rd Angle is the default.

6. To begin writing the program, access the Simulate menu and select the Edit option.

7. Once you are in the text editor, type in the program name. Use one of the examples that are installed in the Demomill or Demoturn folders to get a feel for how interactive graphics simulation works. Note how each line of code is simulated when entered and how the status windows are also updated.

   If there is an error in the code, use the cursor keys to return to the flawed line; then edit it and press Enter. The program will resimulate the block lines from the beginning of the CNC file to the modified block code line.

8. To exit the interactive graphics editor, press F3 as indicated at the bottom of the screen.

9. To watch the entire simulation, exit Edit (press F3) and then access the Simulate menu and select the Cycle Start option. The program currently in memory will be cycled through once. Remember to watch for user input prompts (for example, tool changes).

SAVING A NEW CNC FILE

To save the current CNC file, access the File menu; then select the Save... option. If the file does not have a name, you will be prompted to enter one. If the file name has been previously saved, you will be prompted to decide whether you want to overwrite it. Here you may enter a different name.

PRINTING THE CURRENT CNC PROGRAM FILE

To get a hard-copy list of the current CNC program, access the File menu and choose the Print option. The Windows Print dialog will then be displayed wherein you can select the Windows printer to print to. Note that for all CNCez printing, the default is Portrait mode.

EXITING THE SIMULATOR

To quit or exit the simulator, access the File menu and select the Exit option. If the current file has not been saved, you will be prompted to save it. Selecting Exit returns you to the initial operating system or menu program that launched the simulator.

CHAPTER 5

CNC Milling

CHAPTER OBJECTIVES

After studying this chapter, the student should have knowledge of the following:

Programming for milling operations

Linear and circular interpolation programming

Cutter diameter compensation

Letter address commands for the CNC Mill

When working through Chapters 5 and 6, have the simulation software running on your computer and enter the programs, using either the Simulate Edit or the MDI option.

When simulating programming codes or creating new programs, follow these steps:

1. Select New from the File menu. Enter a file name.
2. Change the Tool Library if required.
3. Select the Workpiece menu. Specify or load a new workpiece.
4. Select the Simulate menu.

Use Edit if you want to create a program and run it or MDI simply to test a code.

Note: The completed sample programs are in the *Demomill* folder.

LETTER ADDRESS LISTING

Letter addresses are variables used in the G- and M-codes. Most G-codes contain a variable, defined by the programmer, for each specific function. Each letter used in conjunction with G-codes or M-codes is called a word. The following letters are used for programming:

D	Diameter offset register number
F	Assigns a feedrate
G	Preparatory function
H	Height offset register number
I	X axis incremental location of arc center
J	Y axis incremental location of arc center
K	Z axis incremental location of arc center
M	Miscellaneous function
N	Block number (specifies the start of a block)
P	Dwell time
R	Retract distance used with G81, 82, 83
	Radius when used with G02 or G03
S	Sets the spindle speed
T	Specifies the tool to be used
X	X axis coordinate
Y	Y axis coordinate
Z	Z axis coordinate

The specific letter addresses are described in more detail next.

Character	Address For
D	Offset register number. Used to call the specified offset register for cutter diameter compensation.
F	Feedrate function. Specifies a feedrate in inches per minute or millimeters per minute.

G	Preparatory function. Specifies a preparatory function. Allows for various modes (for example, rapid and feed) to be set during a program.
H	Offset register number. Used to call the specified offset register for cutter tool length compensation.
I	Circular interpolation. Used in circular motion commands (see G02 and G03) to specify X incremental distance from the startpoint to the centerpoint of the arc.
J	Circular interpolation. Used in circular motion commands (see G02 and G03) to specify Y incremental distance from the startpoint to the centerpoint of the arc.
K	Circular interpolation. Used in circular motion commands (see G02 and G03) to specify Z incremental distance and direction from the startpoint to the centerpoint of the arc.
M	Miscellaneous function. Programmable on/off switches for various machine tool functions, as covered in the next section.
N	Block number. Used for program line identification. Allows the programmer to organize each line and is helpful during editing. CNCez increments block numbers by a selected amount to allow extra lines to be inserted if needed during editing.
P	Dwell time. Used to specify the length of time in seconds in a dwell command (see G04).
R	Retract distance. The Z retract distance in drilling operations. Radius, when used with G02 or G03. Can also be used in circular movement commands (see G02 and G03) to provide an easier way to designate the radius of the circular movement.
S	Spindle speed function. Specifies the spindle speed in revolutions per minute.
T	Tool number select function. Specifies the turret position of the current tool.
X	X-axis definition. Designates a coordinate along the X axis.
Y	Y-axis definition. Designates a coordinate along the Y axis.
Z	Z-axis definition. Designates a coordinate along the Z axis.

G-CODES

G-codes are preparatory functions that involve actual tool moves (for example, control of the machine). These include rapid moves, feed moves, radial feed moves, dwells, and roughing and profiling cycles. Most G-codes described here are modal, meaning that they remain active until canceled by another G-code. The following codes are described in more detail in the following sections.

G00	Positioning in rapid	Modal
G01	Linear interpolation	Modal
G02	Circular interpolation (CW)	Modal
G03	Circular interpolation (CCW)	Modal
G04	Dwell	
G17	XY plane	Modal
G18	XZ plane	Modal
G19	YZ plane	Modal
G20/G70	Inch units	Modal
G21/G71	Metric units	Modal
G28	Automatic return to reference point	
G29	Automatic return from reference point	
G40	Cutter compensation cancel	Modal
G41	Cutter compensation left	Modal
G42	Cutter compensation right	Modal
G43	Tool length compensation (plus)	Modal
G44	Tool length compensation (minus)	Modal
G49	Tool length compensation cancel	Modal
G54-G59	Workpiece coordinate settings	Modal
G73	High-speed peck drilling	Modal
G80	Cancel canned cycles	Modal
G81	Drilling cycle	Modal
G82	Counter boring cycle	Modal
G83	Deep hole drilling cycle	Modal
G90	Absolute positioning	Modal
G91	Incremental positioning	Modal
G92	Reposition origin point	
G98	Set initial plane default	
G99	Return to retract (rapid) plane	

G00 POSITIONING IN RAPID

Format: N_ G00 X_ Y_ Z._

The G00 command is a rapid tool move. A rapid tool move is used to move the tool linearly from position to position without cutting any material. This command is not to be used for cutting any material, as to do so would seriously damage the tool and ruin the workpiece. This command is modal.

On most CNC machine tools, it is standard to program a G00 rapid for an XY move only and the Z moves separately. See Figs. 5.1, 5.2, and 5.3.

EXAMPLE: N25 G00 X2.5 Y4.75   (Rapid to X2.5,Y4.75)

           N30 Z0.1   (Rapid down to Z0.1)

Depending on where the tool is located, there are two basic rules to follow for safety's sake:

1. If the Z value represents a cutting move in the negative direction, the X and Y axes should be executed first.
2. If the Z value represents a move in the positive direction, the X and Y axes should be executed last.

Sample Program G00EX1:

Workpiece Size: X6,Y4,Z1

Tool: Tool #2, 1/4" Slot Drill

Tool Start Position: X0,Y0,Z1

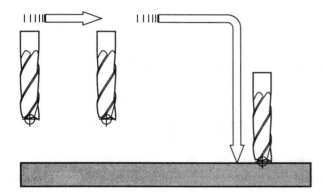

FIGURE 5.1
The G00 command is used to move the tool quickly from one point to another without cutting, thus allowing for quick tool positioning.

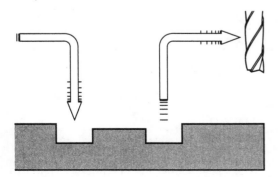

FIGURE 5.2
Note that the G00 rapid move should have two distinct movements to ensure that vertical moves are always separate from horizontal moves. In a typical rapid move toward the part, the tool first rapids in the flat, horizontal XY plane. Then, it feeds down in the Z axis. When rapiding out of a part, the G00 command always goes up in the Z axis first, then laterally in the XY plane.

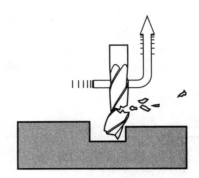

FIGURE 5.3
As this diagram shows, if the
basic rules are not followed, an
accident can result. Improper use
of G00 often occurs because
clamps are not taken into consideration. Following the basic rules
will reduce any chance of error.

For this example it may be helpful to toggle on the Showpaths option
(under the Options Menu).

%	(Program start flag)
:1001	(Program number 1001)
N5 G90 G20	(Absolute and inch programming)
N10 M06 T2	(Tool change, Tool #2)
N15 M03 S1200	(Spindle on CW, at 1200 rpm)
N20 G00 X1 Y1	(Rapid over to X1,Y1)
N25 Z0.1	(Rapid down to Z0.1)
N30 G01 Z-0.25 F5	(Feed move down to a depth of 0.25 in.)
N35 Y3	(Feed move to Y3)
N40 X5	(Feed to X5)
N45 X1 Y1 Z-0.125	(Feed to X1,Y1,Z–0.125)
N50 G00 Z1	(Rapid up to Z1)
N55 X0 Y0	(Rapid over to X0,Y0)
N60 M05	(Spindle off)
N65 M30	(End of program)

Note how in the first rapid section, N20, N25, the tool first moves in the
horizontal plane and then down the Z axis. In the second rapid section,
N50,N55, the tool first moved up and then over to (X0, Y0) because the
tool was into the part.

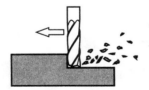

G01 LINEAR INTERPOLATION

Format: N_ G01 X_ Y_ Z_ F_

Linear interpolation is nothing more than straight-line feed moves. A
G01 command is specifically for the linear removal of material from a
workpiece, in any combination of the X, Y, or Z axes.

The G01 is modal and is subject to a user variable feedrate (designated
by the letter F followed by a number). The G01 is not limited to one
plane. Three-axis, simultaneous feed moves, where all three axes are

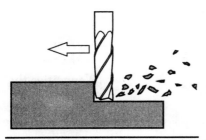

FIGURE 5.4
Linear Interpolation, or straight-line feed moves, on the flat XY plane (no Z values are specified).

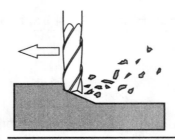

FIGURE 5.5
G01 command, using multiaxis feed moves. All diagonal feed moves are a result of a G01 command, where two or more axes are used at once.

used at the same time to cut different angles, are possible. See Figs. 5.4 and 5.5.

Sample Program G01EX2:

Workpiece Size: X4, Y3, Z1

Tool: Tool #3, 3/8" Slot Drill

Tool Start Position: X0, Y0, Z1

```
%                        (Program start flag)
:1002                    (Program #1002)
N5 G90 G20               (Block #5, absolute in inches)
N10 M06 T3               (Tool change to Tool #3)
N15 M03 S1250            (Spindle on CW at 1250 rpm)
N20 G00 X1.0 Y1.0        (Rapid over to X1,Y1)
N25 Z0.1                 (Rapid down to Z0.1)
N30 G01 Z-0.125 F5       (Feed down to Z-0.125 at 5 ipm)
N35 X3 Y2 F10            (Feed diagonally to X3,Y2 at 10 ipm)
N40 G00 Z1.0             (Rapid up to Z1)
N45 X0.0 Y0.0            (Rapid over to X0,Y0)
N50 M05                  (Spindle off)
N55 M30                  (Program end)
```

In the sample program, several different examples of the G01 command are shown:

- The first G01 command (in N30) instructs the machine to plunge feed the tool below the surface of the part by 0.125 in. at a feedrate of 5 in./min.
- N35 is a two-axis (X and Y) diagonal feed move, and the linear feedrate is increased to 10 ipm.

Note: Because there is contact between the cutting tool and the workpiece, it is imperative that the proper spindle speeds and feedrates be used. It is the programmer's responsibility to ensure acceptable cutter speeds and feeds.

IMPORTANT

G02 CIRCULAR INTERPOLATION (CW)

Format:	N_ G02 X_ Y_ Z_ I_ J_ K_ F_	(I, J, K specify the radius)
or	N_ G02 X_ Y_ Z_ R_ F_	(R specifies the radius)

Circular Interpolation is more commonly known as radial (or arc) feed moves. The G02 command is used specifically for all clockwise radial feed moves, whether they are quadratic arcs, partial arcs, or complete circles, as long as they lie in any one plane. The G02 command is modal and is subject to a user-definable feedrate.

EXAMPLE: G02 X2 Y1 I0 J-1

IMPORTANT

The I and J values represent the relative, or incremental, distance from the startpoint to the arc center.

The G02 command requires an endpoint and a radius in order to cut the arc (see Fig. 5.6). The startpoint of this arc is (X1, Y2) and the endpoint is (X2, Y1). To find the radius, simply measure the relative, (or incremental), distance from the startpoint to the centerpoint. This radius is written in terms of the X and Y distances. To avoid confusion, these values are assigned variables called I and J, respectively.

EXAMPLE: G02 X2 Y1 R1

REMEMBER

You can also specify G02 by entering the X and Y endpoints and then R for the radius.

Note: The use of an R value for the radius of an arc is limited to a maximum movement of 90°.

An easy way to determine the radius values (the I and J values) is by making a small chart:

Centerpoint	X1	Y1
Startpoint	X1	Y2
Radius	I0	J-1

Finding the I and J values is easier than it first seems. Follow these steps:

1. Write the X and Y coordinates of the arc's centerpoint.
2. Below these coordinates, write the X and Y coordinates of the arc's startpoint.
3. Draw a line below this to separate the two areas to perform the subtraction.

Result: G02 X2 Y1 I0 J-1 F5

FIGURE 5.6
Shows arc startpoint, endpoint, and centerpoint.

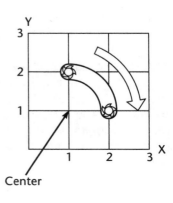

4. To find the I value, calculate the difference between the arc's startpoint and centerpoint in the X direction. In this case, both X values are 1. Hence there is no difference between them, so the I value is 0. To find the J value, calculate the difference between the arc's startpoint and centerpoint in the Y direction. In this case, the difference between Y2 and Y1 is down 1 inch, so the J value is –1.

Sample Program G02EX3:

Workpiece Size: X4, Y3, Z1

Tool: Tool #2, 1/4" Slot Drill

Tool Start Position: X0, Y0, Z1

```
%
:1003
N5 G90 G20
N10 M06 T2
N15 M03 S1200
N20 G00 X1 Y1
N25 Z0.1
N30 G01 Z-0.1 F5
N35 G02 X2 Y2 I1 J0 F20    (Arc feed CW, radius I1,J0 at 20 ipm)
N40 G01 X3.5
N45 G02 X3 Y0.5 R2         (Arc feed CW, radius 2)
N50 X1 Y1 R2               (Arc feed CW, radius 2)
N55 G00 Z0.1
N60 X2 Y1.5
N65 G01 Z-0.25
N70 G02 X2 Y1.5 I0.25 J-0.25 (Full circle arc feed move CW)
N75 G00 Z1
N80 X0 Y0
N85 M05
N90 M30
```

G03 CIRCULAR INTERPOLATION (CCW)

Format: N_ G03 X_ Y_ Z_ I_ J_ K_ F_   (I, J, K specify the radius)

or   N_ G03 X_ Y_ Z_ R_ F_   (R specifies the radius)

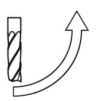

Circular interpolation is more commonly known as radial (or arc) feed moves. The G03 command is specifically for all counterclockwise radial feed moves, whether they are quadratic arcs, partial arcs, or complete circles, as long as they lie in any one plane. The G03 is modal and is subject to a user-definable feedrate.

FIGURE 5.7
Shows arc startpoint, endpoint, and centerpoint.

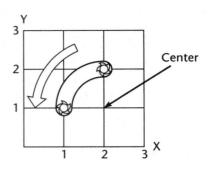

EXAMPLE: G03 X1 Y1 I0 J-1

The G03 command requires an endpoint and a radius in order to cut the arc. (See Fig. 5.7.) The startpoint of this arc is (X2, Y2) and the endpoint is (X1, Y1). To find the radius, simply measure the incremental distance from the startpoint to the centerpoint of the arc. This radius is written in terms of the X and Y distances. To avoid confusion, these values are assigned variables called I and J, respectively.

EXAMPLE: G03 X1 Y1 R1

You can also specify G03 by entering the X and Y endpoints and then R for the radius.

Note: The use of an R value for the radius of an arc is limited to a maximum movement of 90°.

An easy way to determine the radius values (the I and J values) is to make a small chart as follows.

Centerpoint	X2	Y1
Startpoint	X2	Y2
Radius	I0	J-1

Finding the I and J values is easier than it first seems. Follow these steps:

1. Write the X and Y coordinates of the arc's centerpoint.
2. Below these coordinates, write the X and Y coordinates of the arc's startpoint.
3. Draw a line below this to separate the two areas to perform the subtraction.
4. To find the I value, calculate the difference between the arc's startpoint and centerpoint in the X direction. In this case, both X values are 2. Hence there is no difference between them, so the I value is 0.

To find the J value, calculate the difference between the arc's startpoint and centerpoint in the Y direction. In this case, the difference between Y2 and Y1 is down 1 inch, so the J value is −1.

Result: G03 X1 Y1 I0 J-1

Sample Program G03EX4.

Workpiece Size: X4, Y4, Z0.25

Tool: Tool #2, 1/4" Slot Drill

(continues)

(continued)

Tool Start Position: X0, Y0, Z1

```
%
:1004
N5 G90 G20
N10 M06 T2
N15 M03 S1200
N20 G00 X2 Y0.5
N25 Z0.125
N30 G01 Z-0.125 F5
N35 X3 F15
N40 G03 X3.5 Y1 R0.5        (G03 arc using R value)
N45 G01 Y3
N50 G03 X3 Y3.5 I-0.5 J0    (G03 arc using I and J)
N55 G01 X2
N60 G03 X2 Y1.5 I0 J-1      (180° arc using I and J)
N65 G01 Y0.5
N70 G00 Z0.1
N75 X1.5 Y2.5
N80 G01 Z-0.25 F5
N85 G03 X1.5 Y2.5 I0.5 J0   (Full circle using I and J)
N90 G00 Z1
N95 X0 Y0
N100 M05
N105 M30
```

G04 DWELL

Format: N_ G04 P_

The G04 command is a nonmodal Dwell command that halts all axis movement for a specified time, while the spindle continues revolving at the specified rpm (see Fig. 5.8). A Dwell is used largely in drilling operations, which allows for the clearance of chips. Use of the Dwell command is also common after a mill plunge move and prior to starting a linear profile move.

This command requires a specified duration, denoted by the letter P, and followed by the time in seconds.

REMEMBER

It is a good practice to program a Dwell command after a mill plunge move.

FIGURE 5.8
The tool will pause for a short time only, rarely more than several seconds. For a definite program pause, refer to the M00 and M01 commands. Being nonmodal, the G04 must be reentered each time Dwell is to be executed.

Sample Program G04EX5:

Workpiece Size:	X3.5, Y2, Z0.5
Tool:	Tool #1, 1/8" Slot Mill
Tool Start Position:	X0, Y0, Z1

%	(Program start flag)
:1005	(Program #1005)
N5 G90 G20	(Absolute programming in inch mode)
N10 M06 T1	(Tool change to Tool #1)
N15 M03 S1300	(Spindle on CW at 1300 rpm)
N20 G00 X3 Y1 Z0.1	(Rapid to X3,Y1,Z0.1)
N25 G01 Z-0.125 F5.0	(Feed down to Z–0.125 at 5 ipm)
N30 G04 P2	(Dwell for 2 seconds)
N35 G00 X2 Z0.1	(Rapid up to 0.1 and over to X2)
N40 G01 Z-0.125 F5.0	(Feed down to Z–0.125)
N45 G04 P1	(Dwell for 1 second)
N50 G00 Z1.0	(Rapid out to Z1)
N55 X0. Y0.	(Rapid to X0, Y0)
N60 M05	(Spindle off)
N65 M30	(Program end)

G17 XY PLANE

Format: N_ G17

The G17 command sets the system to default to the XY plane as the main machining plane for specifying circular interpolation moves and/or cutter compensation.

On any three-axis machine tool—X, Y, and Z—the tool can move in two basic directions: horizontally (in the X or Y direction) and vertically (in the Z direction). On a simple two-dimensional part (for example, a milled pocket or part profile) the X and Y axes make up the main machining plane, which is horizontal. Here the Z axis is secondary and works perpendicular to the XY plane. G17 is a system default, as it is the most common machining plane. This mode of operation is sometimes referred to as 2-1/2 axis machining. Figure 5.9 shows the G17 plane as the machining plane used for circular arc moves.

Sample Program G17EX6:

Workpiece Size:	X3, Y2, Z1
Tool:	Tool #4, 1/2" Slot Drill
Tool Start Position:	X0, Y0, Z1

%	
:1006	
N5 G90 G20 **G17**	(Set XY plane)
N10 M06 T4	
N15 M03 S1200	
N20 G00 X2 Y1	

(continues)

(continued)

```
N25 Z0.125
N30 G01 Z-0.05 F5
N35 G02 X1 R1 F10
N40 G00 Z1
N45 X0 Y0
N50 M05
N55 M30
```

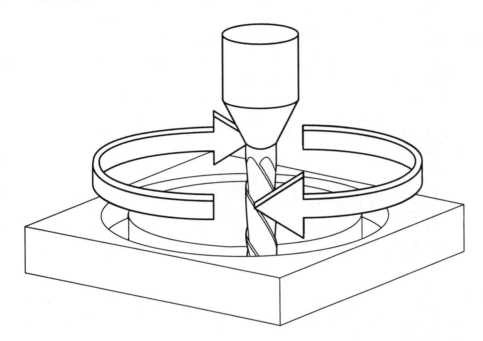

FIGURE 5.9
A circular tool move in the G17 plane.

G18 XZ PLANE

Format: N_ G18

The G18 command sets the system to the XZ plane as the main machining plane for specifying circular interpolation moves and/or cutter compensation.

This command changes the default machining plane to the XZ plane, where the Y axis is secondary, and works perpendicular to the XZ plane. In this plane, it is possible to cut convex or concave arcs using the G02 and G03 circular interpolation commands. See Fig. 5.10. It is important to note that, because the X and Z axes are primary, the radius is no longer expressed in terms of I and J (remember, the distance from the startpoint to the centerpoint; see G02 and G03), but rather in terms of I and K.

Remember, also, to determine the direction of travel look down at the two axes from the Y+ direction in the same way that you look down at the XY axis from the Z+ axis in the G17 plane.

REMEMBER

When programming G02 and G03 commands, keep in mind that the primary and secondary axes are reversed. This means that the G02 will look like a counterclockwise arc and that the G03 will look like a clockwise arc. See the following sample program to better understand this command.

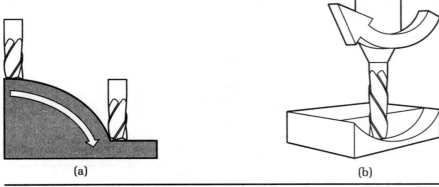

(a)   (b)

FIGURE 5.10

(a) A tool cutting an arc in the XZ plane. (b) An example of an arc cut in the G18 XZ plane. Keep in mind that, because the primary and secondary axes are reversed, this arc is actually a G03 command.

Sample Program G18EX7:

Workpiece Size: X2, Y1, Z1

Tool: Tool #2, 1/4" Slot Drill

Tool Start Position: X0, Y0, Z1

With this example, it is a good idea to toggle on the Showpath option (under the Options menu).

```
%
:1007
N5 G90 G20 G17                 (G17 sets XY plane)
N10 M06 T2
N15 M03 S1200
N20 G00 X0 Y0
N25 Z1
N30 Z0.1
N35 G01 Z0 F5
N40 G18 G02 X2 Z0 I1 K0        (G18 sets XZ plane)
N45 G01 Y0.25
N50 G03 X0.5 Z0 I-0.75 K0
N55 G01 Y0.5 F10
N60 G02 X1.5 Z0 I0.5 K0
N65 G00 Z1
N70 X0 Y0
N75 M05
N80 M30
```

G19 YZ PLANE

Format: N_ G19

The G19 command sets the system to default to the YZ plane as the main machining plane for specifying circular interpolation moves and/or cutter compensation. See Fig. 5.11.

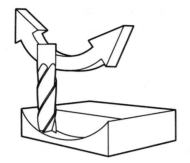

FIGURE 5.11
Tool cutting an arc in the YZ plane.

This command changes the default machining plane to the YZ plane, where the X axis is secondary, and works perpendicular to the YZ plane. In this plane, it is possible to cut convex or concave arcs by using the G02 and G03 circular interpolation commands. It is important to note that because the Y and Z axes are primary, the radius is no longer expressed in terms of I and J (remember, the distance from the startpoint to the centerpoint), but rather in terms of J and K.

Remember, also, to determine the direction of travel, look down at the two axes from the X+ direction in the same way that you look down at the XY axis from the Z+ axis in the G17 plane.

Sample Program G19EX8:

Workpiece Size: X2, Y2, Z1

Tool: Tool #2, 1/4" Slot Mill

Tool Start Position: X0, Y0, Z1

With this example, it is a good idea to toggle on the Showpath option (under the Options menu).

```
%
:1008
N5 G90 G20 G17            (Set XY Plane)
N10 M06 T2
N15 M03 S1200
N20 G00 X0 Y0
N25 Z0.1
N30 G01 Z0 F5
N35 G19 G03 Y1 Z0 J0.5 K0    (Set YZ plane)
N40 G01 X1.5 Y2 F10
N45 G02 Y0 Z0 J-1 K0 F5
N50 G00 Z1
N55 X0 Y0
N60 M05
N65 M30
```

G20 / G70 IMPERIAL INCH UNITS

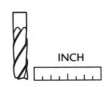

INCH

Format: N_ G20

The G20 command defaults the system to inch data units. When you are running a program and encounter the G20 command, all coordinates are stated in inch units.

This command is the CNCez system default and is modal (see G21).

Sample Program G20EX9:

Workpiece Size: X4, Y2, Z1

Tool: Tool #2, 1/4" Slot Drill

Tool Start Position: X0, Y0, Z1

```
%
:1009
N5 G90 G20                     (Set inch mode)
N10 M06 T2
N15 M03 S1000
N20 G00 X1 Y1
N25 Z1
N30 G01 Z-0.125 F5
N35 X3.625 F15
N40 Y1.75
N45 G00 Z1
N50 X0 Y0
N55 M05
N60 M30
```

In this program the system is using inch units, all coordinates are in inches, and feedrates are expressed as inches per minute.

G21 / G71 METRIC, OR SI UNITS

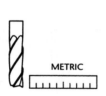

METRIC

Format: N_ G21

The G21 command defaults the system to metric data units (millimeters). After encountering this command, the program states all coordinates in metric (mm) units.

This command is modal (see G20). On most modern controls, it is possible to switch back and forth between metric and inch units within one program. Remember, when using G21 or metric units, you should use the Metric Tool Library.

Sample Program G21EX10:

Workpiece Size: X100, Y75, Z25

Tool: Tool #2, 8mm Slot Drill

Tool Start Position: X0, Y0, Z25.4

Select Metric units under the Setup Menu

```
%
:1010
N5 G90 G21                     (Set metric programming mode)
N10 M06 T2
N15 M03 S1200
N20 G00 X-5 Y5
```

(continues)

(continued)

```
N25  Z-8
N35  G01  X90  F300
N40  X95  Y20
N45  Y50
N50  G03  X75  Y65  R15
N55  G01  X40  Y50
N60  G02  X0  Y10  R40
N65  G00  Z25.4
N70  Y0
N75  M05
N80  M30
```

In this program the system is using metric units, all coordinates are in millimeters, and all feedrates are expressed as millimeters per minute.

G28 AUTOMATIC RETURN TO REFERENCE POINT

Format:　　N_　G28

or　　　　　　N_　G28　X_　Y_　Z_

The G28 command allows the existing tool to be positioned to a predefined reference point automatically via an intermediate position. It can be used prior to programming a tool change command or before a program stop for inspection.

When you are using this command, for safety reasons you should cancel any tool offset or cutter compensation.

All axes are positioned first to the intermediate point at the rapid traverse rate and then from the intermediate point to the reference point (see Fig. 5.12).

The movement from the startpoint to the intermediate point and from the intermediate point to the reference point is the same as for the G00 command.

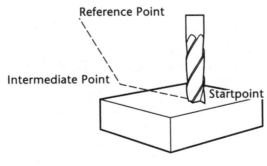

Reference Point

Intermediate Point

Startpoint

FIGURE 5.12
Cutter moves on the G28 command from the startpoint to the intermediate point and finally to the reference point.

Sample Program G28EX111:

Workpiece Size:　　X4, Y4, Z1

Tools:　　　　　　Tool #7, 1" Slot Drill
　　　　　　　　　Tool #10, 1/2" HSS Drill

Tool Start Position: X0, Y0, Z1

Reference Point:　　X0, Y0, Z5

(continues)

(continued)

```
%
:1111
N5  G90 G20
N10 M06 T7
N12 M03 S1000
N15 G00 X4.75 Y2
N20 Z-0.5
N25 G01 X2 F10
N30 G00 Z0.25
N35 G28 X0 Y2.5 Z1        (Return to reference via X0,Y2.5,Z1)
N40 M06 T10
N45 M03 S2000
N50 G29 X2 Y2 Z0.1
N55 G01 Z-1.25 F5
N60 G00 Z1
N65 X0 Y0
N70 M05
N75 M30
```

G29 AUTOMATIC RETURN FROM REFERENCE POINT

Format: N_ G29

or   N_ G29 X_ Y_ Z_

The G29 command can be used immediately after an automatic tool change or program stop for inspection after a G28 command. It allows the tool to be returned to a specified point via the intermediate point, specified by the previous G28 command (see Fig. 5.13).

FIGURE 5.13
Cutter moves on the G29 command from the reference point to the intermediate and finally to the endpoint.

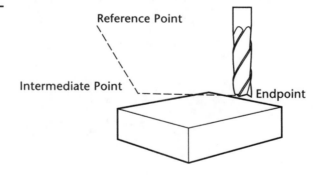

Sample Program G29EX112:

Workpiece Size: X4, Y4, Z1

Tools: Tool #7, 1" Slot Drill
                      Tool #10, 1/2" HSS Drill

Tool Start Position: X0, Y0, Z1

Reference Point: X0, Y0, Z5

```
%
:1112
N5 G90 G20
N10 M06 T7
N12 M03 S1000
N15 G00 X-0.75 Y2
N20 Z-0.5
N25 G01 X-0.5 F10
N30 G03 I2.5 J0
N35 G28 X-0.75 Y2 Z1
N40 M06 T10
N45 M03 S2000
N50 G29 X2 Y2 Z0.1          (Return from reference via X2,Y2,Z0.1)
N55 G01 Z-1.25 F5
N60 G00 Z1
N65 X0 Y0
N70 M05
N75 M30
```

G40 CUTTER COMPENSATION CANCEL

Format: N_ G40

The G40 command cancels any cutter compensation that was applied to the tool during a program and acts as a safeguard to cancel any cutter compensation applied by a previous program.

Cutter compensation is used whenever tool centerline programming is difficult or a certain tool is not available and another tool must be substituted for it. Cutter compensation is also used when excessive tool wear is encountered. Normally, CNC programs are written so that the tool center follows the toolpath. When it is necessary to offset this path either left or right, cutter compensation is used. Remember, cutter compensation is modal, so it must be canceled once it is no longer required. This is the sole function of the G40.

Sample Program G40EX11:

Workpiece Size: X4, Y3, Z1

Tool: Tool #4, 1/2" Slot Drill

Register: D10 is 0.25"

Tool Start Position: X0, Y0, Z1

```
%
:1011
N5 G90 G20 G17 G40          (G40 Compensation cancel)
N10 T04 M06
N15 M03 S1500
N20 G00 X-0.5 Y-0.5
N25 Z-0.5
N30 G01 G42 X0 Y0 D10       (Compensation right)
N35 X3 F10
```

(continues)

(continued)

```
N40 Y3.0
N45 X0
N50 Y0
N55 G00 G40 X-0.5 Y-0.5        (G40 Compensation cancel)
N60 Z1.0
N65 X0 Y0
N70 M05
N75 M30
```

G41 CUTTER COMPENSATION LEFT

Format: N_ G41 D_

The G41 command compensates the cutter a specified distance to the left-hand side of the programmed tool path (see Fig. 5.14). It can be used to compensate for excessive tool wear or to profile a part.

It can also be used to accommodate the lack of a certain specified tool. For example, if a profile was originally programmed for a 1/2" diameter cutter, and for some reason that tool was damaged and no longer available, another tool, say, a 3/8" diameter cutter, could be substituted to achieve the same results. You would just have to modify the offset register values used in the CNC program and enter the proper radius values in the Offset Register Table.

This command specifically refers to the offset registers to determine the correct compensation distance. When you set the offset registers prior to program execution, the MCU is able to refer to them when required. Each numbered register is accessed by the letter address D for CNC milling. To set up the offset registers, refer to Chapter 4 and the Setup menu.

The G41 command is modal, so it compensates each successive tool move the same specified distance, until it is overridden by a G40 command or receives a different offset.

Because the G41 command compensates the tool on the left-hand side of the programmed toolpath, it must first know how long the actual feed move is. During program execution of cutter compensation, a "look-ahead" is performed. By buffering command moves, the program is able

FIGURE 5.14
Cutter positioned on the left-hand side of the cutting line or programmed tool path.

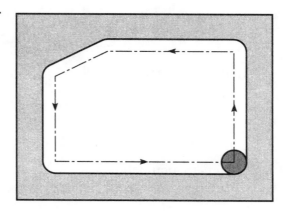

FIGURE 5.15

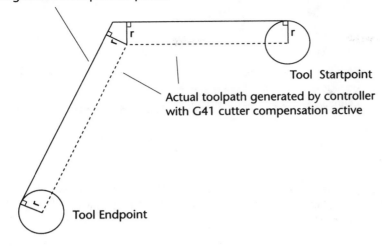

Programmed toolpath on profile

Tool Startpoint

Actual toolpath generated by controller
with G41 cutter compensation active

Tool Endpoint

to determine where the tool currently is located and where it is going. Hence the compensation "lags behind" by one command during program cycles. When a program is executed, the compensation cycle must follow on the left-hand side of the programmed tool path. When the toolpath changes direction, the tool must compensate the appropriate distance from this new direction. The diagram in Fig. 5.15 illustrates the mechanism of cutter radius compensation.

The following program demonstrates the execution of a program with the G41 command. Prior to attempting to run this program, remember to ensure that the offset register table contains the proper entry of .25 for register number 11, as well as set up the tool turret with the appropriate tools from the Tool Library.

Sample program G41EX12:

Workpiece Size:	X5, Y4, Z1
Tool:	Tool #1, 1/4" Slot Drill
	Tool #4, 1/2" End Mill
Register:	D11 is 0.25"
Tool Start Position:	X0, Y0, Z1

```
%
:1012
N5 G90 G20 G40 G17 G80      (Cutter compensation cancel)
N10 T01 M06                 (Tool change to Tool #1)
N15 M03 S2000
N20 G00 X0.5 Y0.5
N25 Z0.1
N30 G01 Z-0.25 F5           (First profile)
N35 X2 F15
N40 X2.5 Y1
N45 Y2
```

(continues)

(continued)

```
N50  G03 X2 Y2.5 R0.5
N55  G01 X0.5
N60  Y0.5                    (End of first profile)
N65  G00 Z1
N70  X0 Y0
N75  T04 M06                 (Tool change to Tool #4)
N80  M03 S1000
N85  G00 X0.75 Y1
N90  Z0.125
N95  G01 Z-0.25 F5           (Second profile begins)
N100 G41 X0.5 Y0.5 D11 F20   (Compensation left)
N105 X2
N110 X2.5 Y1
N115 Y2
N120 G03 X2 Y2.5 R0.5
N125 G01 X0.5
N130 Y0.5
N135 G40 X0.75 Y0.75         (Compensation cancel)
N140 G00 Z1
N145 X0 Y0
N150 M05
N155 M30
```

In this program, the default value for register number 11 is 0.25 in. Note how the G41 works. It is first specified, then the offset register number is referenced, and finally the toolpath is programmed as usual. Sometimes, normal practice is to locate the tool startpoint so the first move is perpendicular to the programmed profile prior to entering a G41 command. See Fig. 5.16.

FIGURE 5.16
The tool entry point is important for profiling a part when cutter radius compensation is being used.

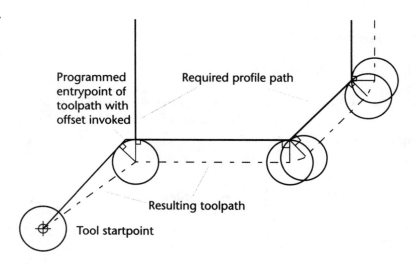

Programmed entrypoint of toolpath with offset invoked

Required profile path

Resulting toolpath

Tool startpoint

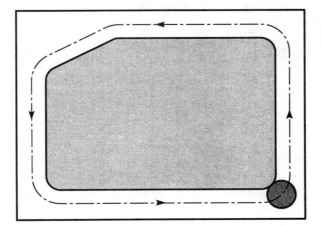

FIGURE 5.17
The cutter positioned on the right-hand side of the programmed toolpath.

G42 CUTTER COMPENSATION RIGHT

Format: N_ G42 D_

In contrast to the G41 command, the G42 command compensates the cutter a specified distance to the right-hand side of the programmed tool path (see Fig. 5.17). It is used to compensate for excessive tool wear or to profile a part. (Sometimes it is easier to compensate the tool than to calculate new arc moves.)

The G42 command refers to the offset registers to determine the correct compensation distance. Setting the offset registers prior to program execution allows the MCU to refer to them when required. Each numbered register is accessed by the letter address D for CNC milling. To set up the offset registers, refer to Chapter 4 and the Setup menu.

The G42 command is modal, so it compensates each successive tool move the same specified distance, until it is overridden by a G40 command or receives a different offset.

Because the G42 command compensates the tool on the right-hand side of the programmed toolpath, it must first know how long the actual feed move is. During program execution of cutter compensation, a "look-ahead" is performed. By buffering command moves the program is able to determine where the tool currently is located and where it is going. Hence the offset "lags behind" by one command during program cycles. When a program is executed, the compensation cycle must follow on the right-hand side of the programmed toolpath. When the toolpath changes direction, the tool must compensate the appropriate distance from this new direction.

The following program illustrates the execution of a program with the G42 command. Prior to attempting to run this program, remember to ensure that the offset register table contains the proper entry of .25 for register number 11, as well as set up the tool turret with the appropriate tools from the Tool Library.

Sample Program G42EX13:

Workpiece Size:	X4, Y4, Z1
Tool:	Tool #1, 1/4" Slot Drill
	Tool #4, 1/2" End Mill
Register:	D11 is 0.25"
Tool Start Position:	X0, Y0, Z1

```
%
:1013
N5 G90 G20 G40 G17 G80      (Setup defaults)
N10 T01 M06                 (Tool change to Tool #1)
N15 M03 S2000
N20 G00 X0.5 Y0.5
N25 Z0.1
N30 G01 Z-0.25 F5           (First profile begins with no comp.)
N35 X2 F15
N40 X2.5 Y1
N45 Y2
N50 G03 X2 Y2.5 R0.5
N55 G01 X0.5
N60 Y0.5
N65 G00 Z1                  (End of first profile)
N70 X0 Y0
N75 T04 M06                 (Tool change to Tool #4)
N80 M03 S1000
N85 G00 X-0.5
N90 Z-0.5
N95 G01 G42 X0.5 Y0.5 Z-0.5 D11 F15   (Second profile with
                                       comp.)
N100 X2
N105 X2.5 Y1
N110 Y2
N115 G03 X2 Y2.5 R0.5
N120 G01 X0.5
N125 Y0
N130 G01 G40 Z0.25          (G40 compensation cancel)
N135 G00 Z1
N140 X0 Y0
N145 M05
N150 M30
```

The default value for register number 11 is 0.25 in. The actual value of D11 does not influence the direction of the compensation (left or right), only the offset distance. Note how the G42 command can be an integral part of a feed move command. The G42 cycle "lags behind" program execution by one block of CNC code so that the tool moves can be calculated.

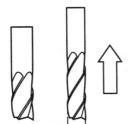

FIGURE 5.18
The G43 command is used when the new tool is longer than the reference tool. The tool must be offset higher so that the endpoints are the same actual height.

G43 TOOL LENGTH COMPENSATION (PLUS)

Format: N_ G43 H_

The G43 command compensates for tool length in a positive direction (see Fig. 5.18). It is important to realize that different tools will have varying lengths, and when tools are changed in a program, any variation in tool length will throw the origin out of zero. To prevent this, the tools can now be compensated for the difference in length for either shorter or longer tools.

This command uses the offset registers found in the Setup menu. The letter address H is used to call up the particular register.

When you are offsetting cutters of different lengths, you must first accurately measure the difference between the two and then enter this value into the offset registers. Failure to measure the tool variation properly will result in unqualified tools and the possibility of machine or workpiece damage or personal injury.

Sample Program G43EX14:

Workpiece Size: X4, Y3, Z1

Tools: Tool #12, 3/8" HSS Drill, 2" length

                      Tool #10, 3/8" HSS Drill, 1.5" length

Tool Start Position: X0, Y0, Z1

Register: 10 is 0.5"

```
%
:1014
N5 G90 G20 G40 G49
N10 M06 T12
N15 M03 S2000
N20 G00 X1 Y1.5
N25 Z0.1
N30 G01 Z-0.5 F5
N35 G00 Z0.1
N40 X2
N45 G01 Z-0.5
N50 G00 Z0.1
N55 X3
```

(continues)

(continued)

```
N60  G01  Z-0.5
N65  G00  Z1
N70  X0  Y0
N75  M06  T10
N80  M03  S1000
N85  G43  H10                    (Cutter compensation 0.25")
N90  G00  X-1.125
N95  Z-0.25
N100 G01  X5.125  F15
N105 G00  Y3
N110 G01  X-1.125
N115 G00  Z1
N120 X0  Y0
N125 G49  M05
N130 M30
```

In this example, register number 13 has a default value of 0.5 in. This means that there is a difference of 0.5 in. between the two tools.

G44 TOOL LENGTH COMPENSATION (MINUS)

Format: N_ G44 H_

The G44 command compensates for tool length in a minus direction (see Fig. 5.19). It is important to realize that different tools will have varying lengths, and when tools are changed in a program, any variation in tool length will throw the origin out of zero. To prevent this, the tools can now be compensated for the difference in length. This command uses the offset registers found in the Setup menu, and the letter address H is used to call a particular register.

Sample Program G44EX15:

Workpiece Size: X4, Y3, Z1

Tools: Tool #4, 1/2" Slot Drill, 1.75" length

                     Tool #12, 1/2" HSS Drill, 2" length

Tool Start Position: X0, Y0, Z1

Register: 11 is 0.25"

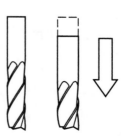

FIGURE 5.19
This is a typical scenario for a G44 command. The new tool is shorter than the original, so the new one must be offset by the difference in their lengths. In this way, the endpoints of both tools are located at the same point in the Z axis.

```
%
:1015
N5 G90 G49 G20
N10 M06 T12
N15 M03 S1200
N20 G00 X1 Y1.5
N25 Z0.25
N30 G01 Z-0.5 F5
N35 G00 Z1
N40 X0 Y0
N45 M06 T4
N50 G44 H11              (Compensate Tool #4—0.25")
N55 G00 X1 Y1.5
N60 Z-0.5
N65 G02 X3 R1.5 F20
N70 X1 R1.5
N75 G01 X3
N80 G00 Z1
N85 X0 Y0
N90 G49 M05
N95 M02
```

G49 TOOL LENGTH COMPENSATION CANCEL

Format: N_ G49

OK

The G49 command cancels all previous cutter length offset commands. Because the G43 and G44 commands are modal, they will remain active until canceled by the G49 command. It is important to keep this in mind; otherwise you may forget that a tool has been offset and crash the cutter into the workpiece.

When cycling programs contain cutter length offsets, it is a good idea to include a G49 command in the program setup, as well as a G49 command to cancel the offsets when they are no longer required.

Sample Program G49EX16:

Workpiece Size: X4, Y3, Z1

Tool: Tool #4, 1/2" Slot Drill, 1.75" length

                      Tool #12, 1/2" HSS Drill, 2" length

Tool Start Position: X0, Y0, Z1

Register: Number 10, 0.25"

```
%
:1016
N5 G90 G49 G20           (Cutter compensation cancel)
N10 M06 T12
```

(continues)

(continued)

```
N15 M03 S1200
N20 G00 X1 Y1.5
N25 Z0.25
N30 G01 Z-0.5 F5
N35 G00 Z1
N40 X0 Y0
N45 M06 T4
N50 G44 H10              (Compensate Tool #4—0.25")
N55 G00 X1 Y1.5
N60 Z-0.5
N65 G02 X3 R1.5 F20
N70 X1 R1.5
N75 G01 X3
N80 G00 Z1
N85 X0 Y0
N90 G49 M05              (Cutter compensation cancel)
N95 M02
```

G54–G59 WORKPIECE COORDINATE SYSTEMS

Format: N_ G54 – G59

The G54–G59 commands are used to establish one of six prepro-grammed work coordinate systems. These settings reside in special para-meter registers in the controller or MCU. Each register has separate X, Y, and Z coordinate settings. The use of these commands can be thought of as special G92 commands for specific work areas. They are frequently used when multiple part fixtures are used in a job, where each register can refer to a specific work area. See Fig 5.20.

Sample Program G54EX19:

Workpiece Size:	X8, Y5, Z2
Tool:	Tool #6, 3/4" HSS Drill
Tool Start Position:	X0, Y0, Z1
Workpiece Coordinate system 2:	X1, Y1, Z0
Workpiece Coordinate system 3:	X5, Y1, Z0

FIGURE 5.20

```
%
:1019
N5 G90 G80 G20
N10 M06 T6
N15 M03 S1300
N20 G55 G00 X1.0 Y1.0      (Rapid to X1, Y1 of work coordinate
                            system 2)
N25 Z0.5
N30 G82 Z-0.25 R0.125 P1 F5
N35 Y2
N40 X2
N45 Y1
N50 X1.5 Y1.5
N60 G80 G00 Z1
N65 G56 G00 X1.0 Y1.0      (Rapid to X1, Y1 of work coordinate
                            system 3)
N70 Z0.5
N75 G82 Z-0.25 R0.125 P1 F5
N80 Y2
N85 X2
N90 Y1
N95 X1.5 Y1.5
N100 G80 G00 Z1
N105 X0 Y0
N110 M05
N115 M30
```

G73 HIGH-SPEED DEEP HOLE DRILLING CYCLE

Format: N_ G73 Z_ R_ Q_ F_

The G73 command involves individual peck moves in each drilling operation. When this command is implemented, the tool positions itself as in a standard G81 drill cycle. The peck is the only action that distinguishes the deep hole drilling cycle from the G81 cycle.

When pecking, the tool feeds in the specified distance (peck distance or depth of cut as specified by the letter address Q, followed by the incremental depth of cut), then rapids up a small predetermined distance. This allows for chip breaking. Because the tool does not rapid back out to the Z retract plane, the drilling process is much faster than the standard G83 peck drilling cycle. The next peck then takes the tool deeper, and it rapids out of the hole the small predefined distance. This process is repeated until the final Z depth is reached. Remember, Q is the incremental depth of cut.

The following sample program demonstrates the G73 command.

Sample Program G73EX20:

Workpiece Size: X4, Y3, Z1

Tool: Tool #3, 3/8" HSS Drill

Tool Start Position: X0, Y0, Z1

(continues)

(continued)

```
%
:1020
N5 G90 G80 G20
N10 M06 T3
N15 M03 S1200
N20 G00 X1 Y1
N25 G73 Z-0.75 R0.125 Q0.0625 F5    (Invoke G73 cycle)
N30 X2.0
N35 X3.0
N40 Y2.0
N45 X2.0
N50 X1.0
N55 G80 G00 Z1                       (Canned cycle cancel)
N60 X0 Y0
N65 M05
N70 M30
```

G80 CANCEL CANNED CYCLES

Format: N_ G80

The G80 command cancels all previous canned cycle commands. As the canned cycles are modal (refer to the canned cycles on the following pages), they will remain active until canceled by the G80 command.

Canned cycles for tapping, boring, spot facing, and drilling are all affected by the G80 command.

When you are creating programs containing canned cycles, it is a good idea to include a G80 command in the program setup at the beginning in the safe-start block, as well as after the drill cycle is completed.

Note: On some controllers the G00 also cancels canned cycles. On others, G01, G02, and G03 may cancel canned cycles as well.

Sample Program G80EX17:

Workpiece Size: X4, Y3, Z1

Tool: Tool #5, 5/8" HSS Drill

Tool Start Position: X0, Y0, Z1

```
%
:1017
N5 G90 G80 G20            (Canned cycle cancel)
N10 M06 T5
N15 M03 S1450
N20 G00 X1.0 Y1.0
N25 G81 Z-0.5 R0.125 F10.0
N30 X2.0
N35 X3.0
N40 G80 G00 Z1.0          (Canned cycle cancel)
N45 X0 Y0
N50 M05
N55 M30
```

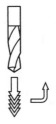

G81 DRILLING CYCLE

Format: N_ G81 X_ Y_ Z_ R_ F_

The G81 command invokes a drill cycle at specified locations. The G81 drill cycle can be used for bolt holes, drilled patterns, and mold sprues, among other tasks. The G81 command is modal and remains active until overridden by another move command or canceled by the G80 command.

The G81 cycle involves several different Z heights:

1. Z initial plane
2. Z depth
3. Z retract plane

The following sample and Figs. 5.21(a)–(c) illustrate how a G81 cycle operates. Remember, the previous Z height before the G81 command (last Z value) is the Z initial plane.

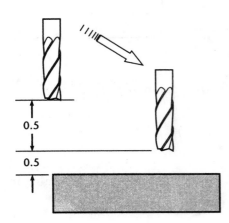

FIGURE 5.21(a)
The first move establishes the Z initial plane. It is the height of the tool before the G81 cycle (0.5 in.). It is a good idea to move the tool from the part to an intermediate level so that less time is wasted rapiding in and out. The tool will rapid back to the initial plane after each hole is drilled.

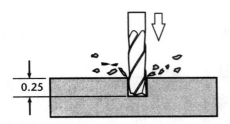

FIGURE 5.21(b)
The tool rapids from the Z initial plane to the retract plane (0.125 in.). Starting at the Z retract plane (0.125 in.), the tool then feeds down to the Z depth (–0.25 in.) at the specified feedrate.

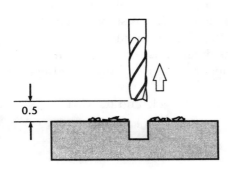

FIGURE 5.21(c)
The tool then rapids up to the Z initial plane (0.5 in.). Then it would go on to the next hole, if any.

In some situations it is advantageous to keep the initial, clearance, and retract planes all at the save level. However, this method is recommended only for experienced programmers.

EXAMPLE: N5 G00 X0 Y0 Z1
N10 X1 Y1 Z0.5
N15 G81 Z-0.25 R0.125 F5

Execute the following sample program to observe the G81 drill cycle. Remember, the G81 command follows a certain sequence.

Sample Program G81EX18:

Workpiece Size: X4, Y3, Z1

Tool: Tool #6, 3/4" HSS Drill

Tool Start Position: X0, Y0, Z1

```
%
:1018
N5 G90 G80 G20
N10 M06 T6
N15 M03 S1300
N20 G00 X1 Y1
N25 Z0.5
N30 G81 Z-0.25 R0.125 F5    (Drill cycle invoked)
N35 X2
N40 X3
N45 Y2
N50 X2
N55 X1
N60 G80 G00 Z1              (Cancel canned cycles)
N65 X0 Y0
N70 M05
N75 M30
N80
```

G82 SPOT DRILLING CYCLE

Format: N_ G82 X_ Y_ Z_ R_ P_ F_

The G82 spot drilling, or counter boring, cycle follows the same operating procedures that the G81 drilling cycle does, with the addition of a dwell. The dwell is a pause during which the Z axis stops moving but the spindle continues rotating (see Fig 5.22). This pause allows for chip clearing and a finer finish on the hole. The dwell time is measured in seconds.

The dwell is initialized by the P letter address, followed by the dwell time in seconds.

The same Z levels apply to both the G82 cycle and the G81 cycle.

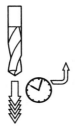

REMEMBER

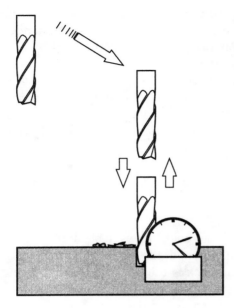

FIGURE 5.22
The diagram shows a dwell at the bottom of the spot drilling cycle. Most CNC controllers have several cycles that utilize a dwell.

Sample Program G82EX19:

Workpiece Size: X4, Y3, Z1

Tool: Tool #6, 3/4" HSS Drill

Tool Start Position: X0, Y0, Z1

```
%
:1019
N5 G90 G80 G20
N10 M06 T6
N15 M03 S1300
N20 G00 X1 Y1
N25 Z0.5
N30 G82 Z-0.25 R0.125 P1.0 F5.0    (Invoke G82)
N35 X2
N40 X3
N45 Y2
N50 X2
N55 X1
N60 G80 G00 Z1
N65 X0 Y0
N70 M05
N75 M30
```

In this program, the following occur:

1. The tool rapids from the Z position to the Z initial plane (see N25).
2. The tool rapids from the Z initial plane (0.5 in.) to the Z retract plane (0.125 in.). This is automatic in the drill cycle.

3. The tool feeds down to the Z depth (–0.25 in.) at the specified feedrate (5 ipm).
4. Then the dwell is executed (P1 specifying a 1-second dwell).
5. The tool returns from the Z depth (–0.25 in.) to the Z retract plane (0.125 in.) at the specified feedrate (F5.0).
6. The tool rapids up to the Z initial plane (0.5 in.), where it is ready to go on to the next hole. To have it rapid to a plane other than the Z initial plane, see the G98 and G99 commands.

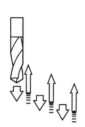

G83 DEEP HOLE DRILLING CYCLE

Format: N_ G83 X_ Y_ Z_ R_ Q_ F_

The G83 command involves individual peck moves in each drilling operation. When this command is implemented, the tool positions itself as in a standard G81 drilling cycle. The peck is the only action that distinguishes the deep hole drilling cycle from the G81 cycle.

When pecking, the tool feeds in the specified distance (peck distance or depth of cut as specified by the letter address Q, followed by the incremental depth of cut), and then rapids back out to the Z retract plane. The next peck takes the tool deeper, and it rapids out of the hole. This process is repeated until the final Z depth is reached. Remember, Q is the incremental depth of cut.

The following sample program demonstrates the G83 command.

Sample Program G83EX20:

Workpiece Size: X4, Y3, Z1

Tool: Tool #3, 3/8" HSS Drill

Tool Start Position: X0, Y0, Z1

```
%
:1020
N5 G90 G80 G20
N10 M06 T3
N15 M03 S1200
N20 G00 X1 Y1
N25 G83 Z-0.75 R0.125 Q0.0625 F5  (Invoke G83 cycle)
N30 X2
N35 X3
N40 Y2
N45 X2
N50 X1
N55 G80 G00 Z1
N60 X0 Y0
N65 M05
N70 M30
```

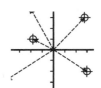

G90 ABSOLUTE POSITIONING

Format: N_ G90

The G90 command defaults the system to accept all coordinates as absolute coordinates. Remember that absolute coordinates are those measured from a fixed origin (X0, Y0, Z0) and are expressed in terms of X , Y, and Z distances.

This command is found at the beginning of most programs to default the system to absolute coordinates. On some machines it is possible to change between absolute and incremental coordinates within a program (see G91).

Remember that the system status window tells you the current coordinate system (see Fig. 5.23).

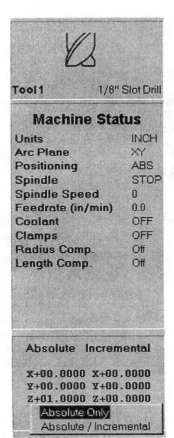

FIGURE 5.23
The Simulator System Status window displays the current coordinate positioning mode.The Simulator Tool Position display can be switched to show both the absolute only or absolute and incremental position by clicking your right mouse button.

Sample Program G90EX21:

Workpiece Size: X4, Y3, Z1

Tool: Tool #2, 1/2" Slot Drill

Tool Start Position: X0, Y0, Z1

(continues)

(continued)

```
%
:1021
N5  G90  G20                    (Set to absolute mode)
N10  M06  T2
N12  M03  S1200
N15  G00  X1  Y1
N20  Z0.125
N25  G01  Z-0.125  F5
N30  X3
N35  Y2
N40  X1
N45  Y1
N50  G00  Z1
N55  X0  Y0
N60  M05
N65  M30
```

G91 INCREMENTAL POSITIONING

Format: N_ G91

The G91 command defaults the system to accept all coordinates as incremental coordinates. Remember that incremental coordinates are measured from the previous point and are expressed in terms of X, Y, and Z distances.

This command is found at the beginning of some programs to default the system to incremental coordinates. It is possible to flip between incremental and absolute coordinates within a program (see G90).

The system status window displays the current coordinate system.

REMEMBER

Sample Program G91EX22:

Workpiece Size: X4, Y3, Z1

Tool: Tool #2, 1/4" Slot Drill

Tool Start Position: X0, Y0, Z1

```
%
:1022
N5  G90  G20
N10  M06  T2
N15  M03  S1200
N20  G00  X1  Y1
N25  Z0.125
N30  G01  Z-0.125  F5
N35  G91  X1  Y1                (Set to incremental mode)
N40  Y-1
N45  X1
N50  Y1
```

(continues)

(continued)

```
N55  G90  G00  Z1
N60  X0  Y0
N65  M05
N70  M30
```

G92 REPOSITION ORIGIN POINT

Format: N_ G92 X_ Y_ Z_

The G92 command is used to reposition the origin point. The origin point is not a physical spot on the machine tool, but rather a reference point to which the coordinates relate. Generally, the origin point is located at a prominent point or object (for example, bottom left and top corner of the part) so that it is easier to measure from.

Sometimes the origin point must be moved. If the operator is to cut several identical parts out of one workpiece, the origin point can be shifted, and the program rerun. Doing this will produce a second part identical to the first, but shifted over from it.

Once you move the origin, it will stay there until you move it back!

REMEMBER

Sample Program G92EX23:

Workpiece Size: X3.5, Y2.5, Z0.75

Tool: Tool #2, 1/4" Slot Drill

Tool Start Position: X0, Y0, Z1

```
%
:1023
N5 G90 G20
N10 M06 T2
N15 M03 S1200
N20 G00 X0.5 Y0.5
N25 Z0.1
N30 G01 Z-0.25 F5
N35 G02 X0.5 Y0.5 I0.25 J0.25 F25
N40 G00 Z0.125
N45 X1.5 Y1.5
N50 G92 X0.5 Y0.5        (Reposition origin)
N55 G01 Z-0.25 F5
N60 G02 X0.5 Y0.5 I0.25 J0.25 F20
N65 G00 Z0.1
N70 X1.5 Y-0.5
N75 G92 X0.5 Y0.5        (Reposition origin)
N80 G01 Z-0.25 F5
N85 G02 X0.5 Y0.5 I0.25 J0.25 F15
N90 G00 Z1
N95 X-2 Y0
N100 G92 X0 Y0          (Reposition origin)
N105 M05
N110 M30
```

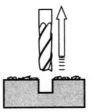

G98 SET INITIAL PLANE RAPID DEFAULT

Format: N_ G98

The G98 command forces the tool to return to the Z initial plane after a drilling operation (see Fig. 5.19). This forces the tool up and out of the workpiece. The G98 return to Z initial plane command is used on workpieces that have clamps or obstacles that could interfere with the tool's movement. The G98 command is also the system default.

Sample Program G98EX24:

Workpiece Size: X3, Y3, Z1

Tool: Tool #3, 3/8" HSS Drill

Tool Start Position: X0, Y0, Z1

```
%
:1024
N5 G90 G80 G20
N10 M06 T3
N15 M03 S1200
N20 G00 X1 Y1
N25 Z0.5
N30 G98 G81 Z-0.25 R0.25 F3      (Set initial plane to Z0.5)
N35 X2
N40 Y2
N45 X1
N50 G80 G00 Z1
N52 X0 Y0
N55 M05
N60 M30
```

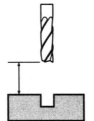

G99 SET RAPID TO RETRACT PLANE

Format: N_ G99

The G99 command forces the tool to return to the retract plane after a drilling operation. This forces the tool up and out of the workpiece, overriding the system default. This command is used on workpieces that do not have surface obstacles. Drilling cycles are quicker when executed, since the tool is moving only to the retract plane. It also is useful for drilling inside a milled pocket. However care must be taken to ensure that the tool clears the pocket at the end of the drilling cycle.

Sample Program G99EX25:

Workpiece Size: X3, Y3, Z1

Tool: Tool #3, 1/2" HSS Drill

Tool Start Position: X0, Y0, Z1

```
%
:1025
N5 G90 G80 G20
```

(continues)

(continued)

```
N10  M06 T3
N15  M03 S1200
N20  G00 X1 Y1
N25  Z0.5
N30  G99 G81 Z-0.25 R0.25 F3  (Set rapid to retract plane)
N35  X2
N40  Y2
N45  X1
N50  G80 G00 Z1
N55  X0 Y0
N60  M05
N65  M30
```

Note: When the tool retracts out of the hole, it stops at the Z retract level. It then rapids to the next X,Y hole location.

IMPORTANT

M-CODES

M-codes are miscellaneous functions that include actions necessary for machining but not those that are actual tool movements. That is, they are auxiliary functions, such as spindle on and off, tool changes, coolant on and off, program stops, and similar related functions. The following codes are described in more detail in the following sections.

M00	Program stop
M01	Optional program stop
M02	Program end
M03	Spindle on clockwise
M04	Spindle on counterclockwise
M05	Spindle stop
M06	Tool change
M08	Coolant on
M09	Coolant off
M10	Clamps on
M11	Clamps off
M30	Program end, reset to start
M98	Call subroutine command
M99	Return from subroutine command
Block Skip	Option to skip blocks that begin with '/'
Comments	Comments may be included in blocks with round brackets '(' ')'

M00 PROGRAM STOP

Format: N_ M00

The M00 command is a temporary program stop function. When it is executed, all functions are temporarily stopped and will not restart unless and until prompted by user input.

The following screen prompt will be displayed with the CNCez simulators: "Program Stop. Enter to Continue." The program will not resume unless and until Enter is pressed. The wording of this prompt varies by machine tool.

This command can be used in lengthy programs to stop the program in order to clear chips, take measurements, or adjust clamps, coolant hoses, and so on.

Sample Program M00EX1:

Workpiece Size: X4, Y3, Z1

Tool: Tool #2, 1/4" Slot Drill

Tool Start Position: X0, Y0, Z1

```
%
:1001
N5 G90 G20
N10 M06 T2
N15 M03 S1200
N20 G00 X1 Y1
N25 Z0.1
N30 G01 Z-0.125 F5
N35 M00                    (Program stop invoked)
N40 G01 X3
N45 G00 Z1
N50 X0 Y0
N55 M05
N60 M30
```

M01 OPTIONAL PROGRAM STOP

Format: N_ M01

The M01 command is an optional stop command and halts program execution only if the Optional Stop switch is set to ON ☑. If the Optional Stop switch is set to OFF ☐, the program will ignore any M01 commands it encounters in a program and no optional stop will be executed.

The M01 optional stop switch is listed in the Options menu and is called Optional Stop. A ☑ beside Optional Stop indicates that it is on. A ☐ indicates that it is off.

The M00 program Stop command is not affected by the Optional Stop switch.

The Optional Stop is used in the following program. Cycle this program once with the switch off ☐, then turn it on ☑, and cycle the program a second time.

Sample Program M01EX2:

Workpiece Size: X4, Y3, Z1

Tool: Tool #2, 1/4" Slot Drill

Tool Start Position: X0, Y0, Z1

Optional Stop: On

```
%
:1002
N5 G90 G20
N10 M06 T2
N15 M03 S1200
N20 G00 X1 Y1
N25 Z0.1
N30 G01 Z-.125 F5
```
N35 M01   (Stop program)
```
N40 G01 X3
N45 G00 Z1
N47 X0 Y0
N50 M05
N55 M30
N60
```

M02 PROGRAM END

Format: N_ M02

The M02 command indicates an end of the main program cycle operation. Upon encountering the M02 command, the MCU switches off all machine operations (for example, spindle, coolant, all axes, and any auxiliaries), terminating the program.

This command appears on the last line of the program.

Sample Program M02EX3:

Workpiece Size: X4, Y3, Z1

Tool: Tool #2, 1/4" Slot Drill

Tool Start Position: X0, Y0, Z1

```
%
:1003
N5 G90 G20
N10 M06 T2
N15 M03 S1200
N20 G00 X1 Y1
N25 Z0.1
N30 G01 Z-.125 F5
N35 X3 F15
N40 G00 Z1
N45 X0 Y0
N50 M05
```
N55 M02   (Program end)

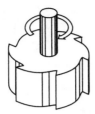

MO3 SPINDLE ON CLOCKWISE

Format: N_ M03 S_

The M03 command switches the spindle on in a clockwise rotation. The spindle speed is designated by the S letter address, followed by the spindle speed in revolutions per minute.

The spindle speed is shown during program simulation in the Program Status window. Its on/off status is shown in the System Status window (CW, CCW, or OFF).

Sample Program M03EX4:

Workpiece Size: X4, Y3, Z1

Tool: Tool #2, 1/4" Slot Drill

Tool Start Position: X0, Y0, Z1

```
%
:1004
N5 G90 G20
N10 M06 T2
N15 M03 S1200          (Spindle on clockwise)
N20 G00 X1 Y1
N25 Z0.25
N30 G01 Z-0.1 F5
N35 X3 F20
N40 X1 Y2 Z-0.5
N45 G19 G02 Y1 Z-0.1 J-0.5 K0.2
N40 G17 G00 Z1
N45 X0 Y0
N50 M05
N55 M30
```

MO4 SPINDLE ON COUNTERCLOCKWISE

Format: N_ M04 S_

The M04 command switches the spindle on in a counterclockwise rotation. The spindle speed is designated by the S letter address, followed by the spindle speed in revolutions per minute.

The spindle speed is shown during program simulation in the Program Status window. Its on/off status is shown in the System Status window (CW, CCW, or OFF).

Sample Program M04EX4:

Workpiece Size: X4, Y3, Z1

Tool: Tool #4, 1/2" Slot Drill

Tool Start Position: X0, Y0, Z1

```
%
:1005
```

(continues)

(continued)

```
N5 G91 G20
N10 M06 T4
N15 M04 S1000          (Spindle on counterclockwise)
N20 G00 X1 Y2
N25 Z-0.75
N30 G01 Z-0.5 F5
N35 X0.5 F20
N40 G03 X0.5 Y0.5 R0.5
N45 X0.5 Y-0.5 R0.5
N50 G01 X0.5
N55 Y-0.25
N60 X-2
N65 Y-0.25
N70 X2
N75 Y-0.25
N80 X-2
N85 Y-0.25
N90 X2
N95 Y-0.25
N100 X-2
N105 G00 Z1.25
N110 G90 X0 Y0
N115 M05
N120 M02
```

M05 SPINDLE STOP

Format: N_ M05

The M05 command turns the spindle off. Although other M-codes turn off all functions (for example, M00 and M01), this command is dedicated to shutting the spindle off directly. The M05 command appears at the end of a program.

Sample Program M05EX5:

Workpiece Size: X4, Y3, Z1

Tool: Tool #2, 1/4" Slot Drill

Tool Start Position: X0, Y0, Z1

```
%
:1005
N5 G90 G20
N10 M06 T2
N15 M03 S1200
N20 G00 X1 Y0.5
N25 Z0.1
```

(continues)

(continued)

```
N30  G01  Z-0.25  F5
N35  G03  X1  Y2.5  I0  J1  F25
N40  X3  I1  J0
N45  Y0.5  I0  J-1
N50  X1  I-1  J0
N55  G00  Z1
N60  X0  Y0
N65  M05                  (Spindle stop)
N70  M30
```

M06 TOOL CHANGE

Format: N_ M06 T_

The M06 command halts all program operations for a tool change. It is actually a two-fold command. First, it stops all machine operations—for example, the spindle is turned off and oriented for the tool change, and all axes motion stops—so that it is safe to change the tool. Second, it actually changes the tool. Look in the Tool Display window; the new tool will have been changed.

For program operation to continue after the tool change, you must respond to the dialog prompt "Toolchange-Press Enter, or click OK to Continue." The program will not continue until you press Enter or click on OK.

Sample Program M06EX6:

Workpiece Size: X4, Y3, Z1

Tool: Tool #8, 3/4" HSS Drill

                   Tool #9, 3/4" End Mill

Tool Start Position: X0, Y0, Z1

```
%
:1006
N5   G90  G20
N10  M06  T8        (Tool change to Tool #8 end mill)
N15  M03  S1000
N20  G00  X0.75  Y1.5
N25  Z0.1
N30  G01  Z-0.5  F2.5
N35  G00  Z0.1
N40  X2.5
N45  G01  Z-0.5
N50  G00  Z1
N55  X0  Y0
N60  M06  T9        (Tool change to Tool #9   3/4" end mill)
N65  X0.75  Y1.5
N70  Z0.1
N75  G01  Z-0.5  F5
```

(continues)

(continued)

```
N80  G02 I0.375 J0 F15
N85  G00 Z0.1
N90  X2.5
N95  G01 Z-0.5 F5
N100 G02 I0.375 J0 F15
N105 G00 Z1
N110 X0 Y0
N115 M05
N120 M30
```

The M06 command halts all functions until Enter is pressed. The new tool is shown in the Simulation and Tool Display windows.

M07/M08 COOLANT ON

Format: N_ M07 or N_ M08

The M07 and M08 commands switch on the coolant flow. Their status is shown in the System Status window.

Sample Program M08EX7:

Workpiece Size: X4, Y3, Z1

Tool: Tool #2, 1/4" Slot Drill

Tool Start Position: X0, Y0, Z1

```
%
:1008
N5  G90 G20
N10 M06 T2
N15 M03 S1200
N20 M08                    (Coolant on)
N25 G00 X1 Y1
N30 Z0.1
N35 G01 Z-.25 F5
N40 X3 F20
N45 Y2
N50 X1
N55 Y1
N60 G00 Z1
N65 M09                    (Coolant off)
N70 G00 X0 Y0
N75 M05
N80 M30
```

M09 COOLANT OFF

Format: N_ M09

The M09 command shuts off the coolant flow. The coolant should be shut off prior to tool changes or when you are rapiding the tool over long distances. Refer to the System Status window to check on the status of the coolant flow.

Sample Program M09EX8:

Workpiece Size: X4, Y3, Z1

Tool: Tool #2, 1/4" Slot Drill

Tool Start Position: X0, Y0, Z1

```
%
:1009
N5 G90 G20
N10 M06 T2
N15 M03 S1200
N20 M08                (Coolant on)
N25 G00 X1 Y1
N30 Z0.1
N35 G01 Z-.25 F5
N40 X3 F20
N45 Y2
N50 X1
N55 Y1
N60 G00 Z1
N65 M09                (Coolant off)
N70 G00 X0 Y0
N75 M05
N80 M30
```

M10 CLAMPS ON

Format: N_ M10

The M10 command turns on the automatic clamps to secure the workpiece. Automatic clamps can be pneumatic, hydraulic, or electromechanical. Not all CNC machines have automatic clamps, but the option exists and the actual code will vary by machine tool make and model.

This command is normally in the program setup section of a CNC program. The System Status window shows the status of the clamps.

Sample Program M10EX9:

Workpiece Size: X4, Y3, Z1

Tool: Tool #12, 1" End Mill

Tool Start Position: X0, Y0, Z1

```
%
:1010
N5 G90 G20
N10 M06 T12
N15 M10                (Clamp workpiece)
```

(continues)

(continued)

```
N20  M03  S1000
N25  G00  X-0.75  Y1
N30  Z-0.375
N35  G01  X0  F10
N40  G03  Y2  I0  J0.5
N45  G01  X2  Y3
N50  X4  Y2
N55  G03  Y1  I0  J-0.5
N60  G01  X2  Y0
N65  X0  Y1
N70  G00  Z1
N75  X0  Y0
N80  M05
N85  M11                    (Unclamp workpiece)
N90  M30
```

M11 CLAMPS OFF

Format: N_ M11

The M11 command releases the automatic clamps so that the workpiece may be removed and the next blank inserted. The automatic clamps may be pneumatic, hydraulic, or electromechanical, depending on the application. The System Status window shows the status of the clamps.

Sample Program M11EX10:

Workpiece Size: X4, Y3, Z1

Tool: Tool #12, 1" End Mill

Tool Start Position: X0, Y0, Z1

```
%
:1011
N5   G90  G20
N10  M06  T12
N15  M10                    (Clamp workpiece)
N20  M03  S1000
N25  G00  X-0.75  Y1
N30  Z-0.375
N35  G01  X0  F10
N40  G03  Y2  I0  J0.5
N45  G01  X2  Y3
N50  X4  Y2
N55  G03  Y1  I0  J-0.5
N60  G01  X2  Y0
N65  X0  Y1
N70  G00  Z1
```

(continues)

(continued)

```
N75 X0 Y0
N80 M05
N85 M11                    (Unclamp workpiece)
N90 M30
```

M30 PROGRAM END, RESET TO START

Format:　　N_ M30

The M30 command indicates the end of the program data. In other words, no more program commands follow it. This is a remnant of the older NC machines, which could not differentiate between one program and the next, so an End of Data command was developed. Now the M30 is used to end the program and reset it to the start.

Sample Program M30EX13:

Workpiece Size:　　　X4, Y3, Z1

Tool:　　　　　　　　Tool #2, 1/4" Slot Drill

Tool Start Position: X0, Y0, Z1

```
%
:1012
N5 G90 G20
N10 M06 T2
N15 M03 S1200
N20 G00 X0.5 Y1.25
N25 Z0.1
N30 G01 Z-0.25 F5
N35 G91 G02 X0.5 Y-0.5 R0.5 F15
N40 X0.5 Y-0.5 R0.5
N45 X1 I0.5
N50 X0.5 Y0.5 I0.5
N55 X0.5 Y0.5 I0.5
N60 G03 X-3 I-1.5
N60 G00 Z1
N65 X0 Y0
N70 M05
N75 M30                    (Program end; reset to start)
```

M98 CALL SUBPROGRAM

Format:　　N_ M98 P_

The M98 function is used to call a subroutine or subprogram. Execution is halted in the main program and started on the program referenced by the P letter address value. For example, N15 M98 P1003 would call program :1003, either from within the current CNC program file or from an external CNC program file. Machine status is maintained when a sub-program is called. This is especially useful in family parts programming or when several operations are required on the same hole locations.

In the following sample program the subprogram is used to drill a hole pattern, using several calls to different drill cycles. The main program positions the machine tool at the starting location to invoke the cycle; the subprogram then continues the pattern.

Sample Program M98EX9:

Workpiece Size: X5, Y5, Z1

Tool: Tool #1, 3/32" Spot Drill

Tool #2, 1/4" HSS Drill

Tool #3, 1/2" HSS Drill

Tool Start Position: X0, Y0, Z1

```
%
:1010
N5  G90 G20
N10 M06 T1
N15 M03 S1500
N20 M08                          (Coolant on)
N25 G00 X1 Y1
N30 G82 X1 Y1 Z-.1 R.1 P0.5 F5  (Start of cycle)
N35 M98 P1005                    (Call subprogram to do rest)
N40 G80
N45 G28 X1 Y1
N50 M09
N55 M06 T02
N60 G29 X1 Y1
N65 M03 S1200
N70 M08
N75 G83 X1 Y1 Z-1 R0.1 Q0.1 F5.0  (Start of cycle)
N80 M98 P1005                    (Call subprogram to do rest)
N85 G80
N90 G28 X1 Y1
N95 M09
N100 M06 T03
N105 G29 X1 Y1
N110 M03 S1000
N115 M08
N120 G73 X1 Y1 Z-1 R0.1 Q0.1 F5.0  (Start of cycle)
N125 M98 P1005                   (Call subprogram to do rest)
N130 G80
N135 G00 Z1
N140 X0 Y0
```

(continues)

(continued)

```
N145 M09
N150 M05
N155 M30
O1005                                    (Subprogram)
N5  X2
N10 X3
N15 X4
N20 Y2
N25 X3
N30 X2
N35 X1
N40 M99                                  (Return from subprogram)
```

M99 RETURN FROM SUBPROGRAM

Format: N_ M99

The M99 function is used to end or terminate the subprogram and return to the main calling program. Execution is continued at the line immediately following the subprogram call. It is used only at the end of the subprogram.

Sample Program M99EX10:

Workpiece Size: X5, Y5, Z1

Tool: Tool #1, 3/32" Spot Drill

                     Tool #2, 1/4" HSS Drill

Tool Start Position: X0, Y0, Z1

```
%
:1011
N5  G90 G20
N10 M06 T1
N15 M03 S1500
N20 M08                              (Coolant on)
N25 G00 X1 Y1
N30 G82 X1 Y1 Z-.1 R.1 P0.5 F5       (Start of cycle)
N35 M98 P1005                        (Call subprogram to do rest)
N40 G80
N45 G28 X1 Y1
N50 M09
N55 M06 T03
N60 G29 X1 Y1
N65 M03 S1200
N70 M08
```

(continues)

(continued)

```
N75 G83 X1 Y1 Z-1 R0.1 Q0.1 F5.0   (Start of cycle)
N80 M98 P1006                       (Call subprogram to do
                                     rest)

N85 G80
N135 G00 Z1
N140 X0 Y0
N145 M09
N150 M05
N155 M30
O1006                               (Subprogram to drill rest of
                                     square pattern)

N5 X2
N20 Y2
N25 X1
N30 M99                             (Return from subprogram)
```

BLOCK SKIP

Format: / N_

The use of the block skip function is extremely helpful in family of parts programming. This functionality is directly controlled by the Block Skip switch on most CNC Controllers. With CNCez simulators this option is found in the Options menu. A checkmark ☑ next to this option item indicates that Block Skip is selected, or turned on. If turned on, upon execution of a CNC program and encountering a "/", the program will ignore any CNC code on that block.

An example of using the Block Skip switch would be when two customers want a similar part machined but only one may need a particular set of operations. For example, one may require a product identification number machined into the part, while the other may not.

When running the following sample program, first try it with the Block Skip on. Then repeat the program with the Block Skip off.

Sample Program SKIPEX1:

Workpiece Size: X12, Y12, Z1

Tool: Tool #1, 3/32" Spot Drill

                   Tool #2, 1/4" HSS Drill

                   Tool #3, 1/2" HSS Drill

Tool Start Position: X0, Y0, Z1

```
%
:1010
N5 G90 G20
N10 M06 T1
N15 M03 S1500
```

(continues)

(continued)

```
N20  M08                                    (Coolant on)
N25  G00  X1  Y1
N30  G82  X1  Y1  Z-.1  R.1  P0.5  F5  (Start of cycle)
N35  M98  P1005                             (Call subprogram to do rest)
N40  G80
N45  G28  X1  Y1
N50  M09
N55  M06  T02
N60  G29  X1  Y1
N65  M03  S1200
N70  M08
N75  G83  X1  Y1  Z-1  R0.1  Q0.1  F5.0  (Start of cycle)
N80  M98  P1005                             (Call subprogram to do rest)
/N85   G80
/N90   G28  X1  Y1
/N95   M09
/N100  M06  T03
/N105  G29  X1  Y1
/N110  M03  S1000
/N115  M08
/N120  G73  X1  Y1  Z-1  R0.1  Q0.1  F5.0  (Start of cycle)
/N125  M98  P1005                             (Call subprogram to do rest)
N130 G80
N135 G00  Z1
N140 X0  Y0
N145 M09
N150 M05
N155 M30

%
:1006                                       (Subprogram to drill square
                                             pattern)
N5   X2
N20  Y2
N25  X1
N30  M99                                    (Return from subprogram)
```

COMMENTS

Format: N_ (Comment statement)

Comments help the CNC machine operator to set up and run a job. Comments are defined by the use of round brackets. Anything between them is ignored by the controller. Throughout the sample programs comments were used to help explain the CNC codes. Remember that comments are just aids to help in reading and understanding a program. Their text is totally ignored even if it contains valid CNC code.

EXAMPLE: N125 G00 Z.5   (RAPID to clearance plane)

STEP-BY-STEP MILLING EXAMPLES

Work through each of the following examples. In the first example, each step is described in more detail than it is in the remaining examples, so be sure to work through it first.

If you have any difficulties with the programs or simply want to test and see them without entering them, you can find them in the */cnc-work/demomill* directory.

For each of the following examples, assume that the PRZ is located at the lower left-hand corner at the top surface of the workpiece. When you have entered the programs in CNCez, the tool will start at (X0, Y0, Z1) relative to the PRZ.

EXAMPLE 1: I-part1.mil

This program introduces you to the Cartesian coordinate system and absolute coordinates. Only single-axis, linear-feed moves show the travel directions of the X, Y, and Z axes. The completed part is shown in Fig. 5.24.

Workpiece Size: X5, Y4, Z1

Tool: Tool #3, 3/8" End Mill

Tool Start Position: X0, Y0, Z1 (Relative to workpiece)

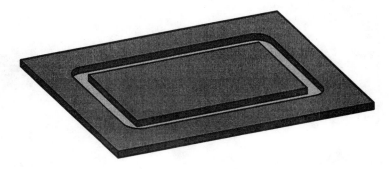

FIG. 5.24
The finished part: a 5 in. x 4 in. block with a 4 in. x 3 in. slot.

```
%
:1001
N5 G90 G20
N10 M06 T3
N15 M03 S1200
N20 G00 X1 Y1
```

(continues)

(continued)

```
N25 Z0.125
N30 G01 Z-0.125 F5
N35 X4 F20
N40 Y3
N45 X1
N50 Y1
N55 G00 Z1
N60 X0 Y0
N65 M05
N70 M30
```

STEP 1: Run the CNCez Simulator either from the Windows taskbar, from the desktop icon, or from the CNC Workshop CBT.

STEP 2: The initial CNCez opening screen will always display the workpiece in 3rd angle projection mode. This mode can be easily changed by right-clicking in the graphics window.

STEP 3: The CNCez graphic user interface is displayed (Fig. 5.25).

STEP 4: Create a new file called I-part1.

Move the pointer to the menu bar and select File.

Select New. (See Fig. 5.26)

A dialog is displayed prompting you to enter the new filename. You will note that a default file extension is

FIGURE 5.25

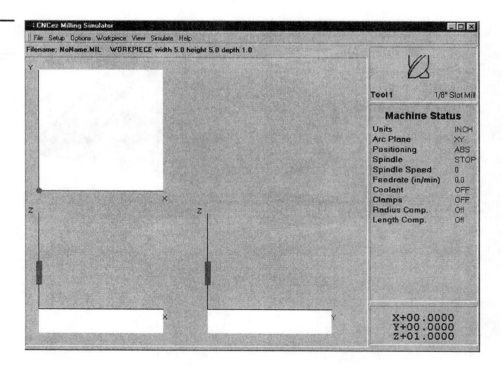

FIGURE 5.26

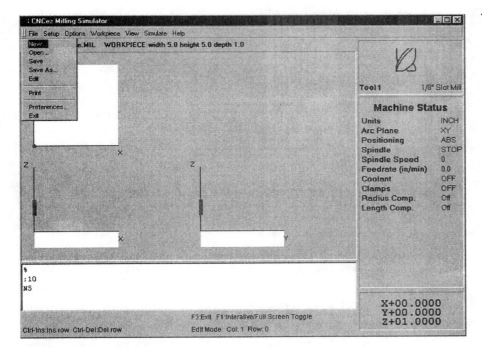

automatically included in the input field. This extension is the default that is set in the Machine Configuration Dialog of the Preferences... option of the File Menu.

The program automatically fills the first three lines of CNC code in the edit box to include the "%" program start code, a program number with :10, and a blank CNC block code N5. Exit this mode for now by using the F3 key.

STEP 5: Set up the workpiece (stock material) for this program.

From the menu bar, select Workpiece.

Select New... (Fig. 5.27).

FIGURE 5.27

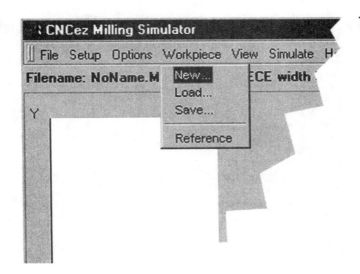

FIGURE 5.28

In the Workpiece Def. dialog enter the workpiece length (in.):
5 (Fig. 5.28).

Enter the workpiece width (in.): 4

Enter the workpiece height (in.): 1

Click on OK.

The workpiece appears in the graphics simulation window in 3rd angle projection, showing the top view, front view, and right-hand side view (Fig. 5.29). You may change views by using the View menu options.

STEP 6: This CNC program will use the default Tool Library that is shipped with the software. The program uses Tool #3, which is defaulted in the turret to be a 3/8" slot drill. Look in the Tool Library at Tool #3.

From the Menu Bar, select Setup.

Select Tools.

Move the pointer over Tool #3 in the Tool Library and click on it. The green box will change to red, indicating that it has been selected (Fig. 5.30).

Now move the pointer over the #3 entry in the Tool Turret table and click on it. You have now set the Tool Turret index #3 with Tool #3 from the Tool Library.

STEP 7: Begin entering the program and simulate the cutter path. Use the Edit option in Simulate.

From the Menu Bar, select Simulate. Then select Edit (Fig. 5.31).

FIGURE 5.29

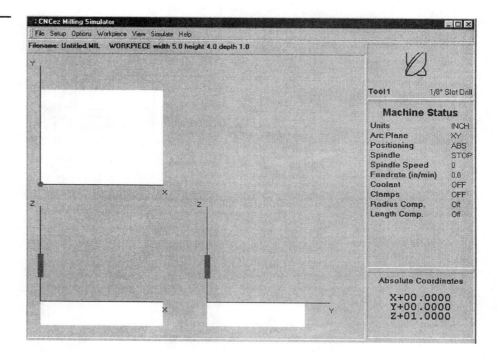

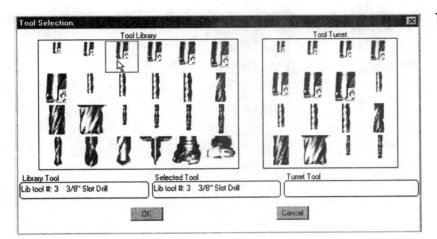

FIGURE 5.30

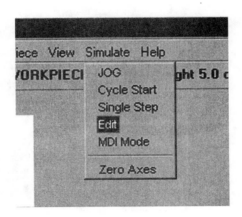

FIGURE 5.31

STEP 8: Program setup phase. You must enter all the setup parameters before you can enter the actual cutting moves (Figs. 5.32, 5.33, and 5.34).

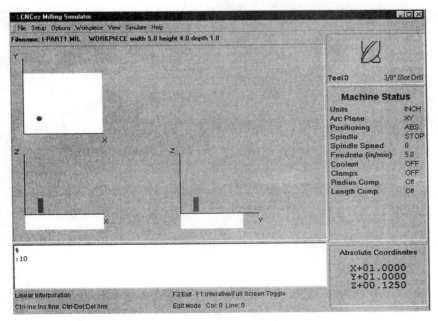

FIGURE 5.32

FIGURE 5.33

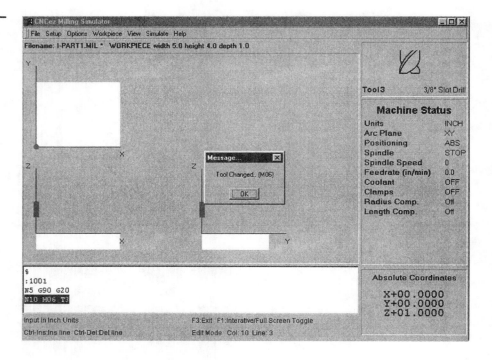

FIGURE 5.34

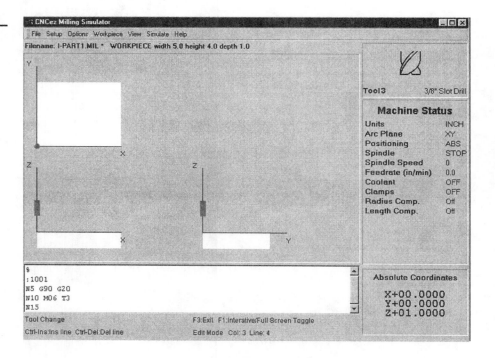

>% [Enter]	Program start flag
>:1001 [Enter]	Program number
>N5 G90 G20 [Enter]	Absolute coordinates and inch
>N10 M06 T3 [Enter]	Tool change and Tool #3 (Fig. 5.33)
>N15 M03 S1200 [Enter]	Spindle on CW at 1200 rpm

STEP 9:   Material removal phase. Begin cutting the workpiece using G00 and G01.

>N20 G00 X1 Y1 [Enter]   Rapid move to (X1, Y1)
>N25 Z0.125 [Enter]   Rapid move to Z0.125 (Fig. 5.35)

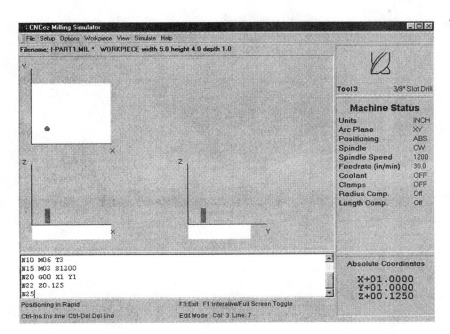

FIGURE 5.35

>N25 G01 Z-0.125 F5 [Enter]   Feed into part 0.125 at 5 ipm (Fig. 5.36)

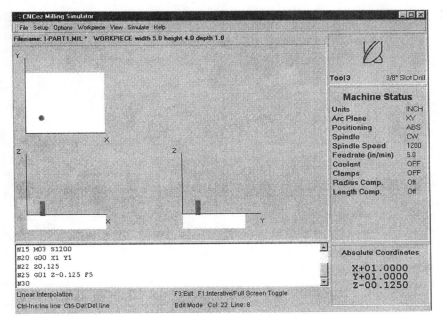

FIGURE 5.36

FIGURE 5.37

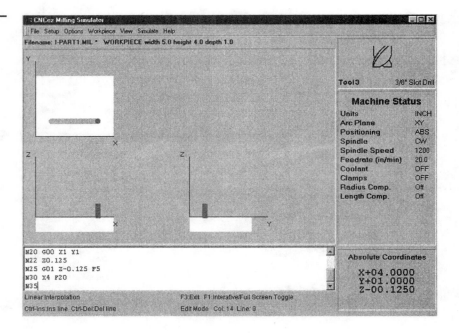

>N30 X4 F20 [Enter]   Feed across to 4 in. at 20 ipm (Fig. 5.37)

Note the status area and the (X, Y, Z) machine coordinate display.

>N40 Y3 [Enter]   Feed up 3 in. (Fig. 5.38)

Note in the status window that the feedrate remains the same. Also note that the G01 is modal and remains in effect until canceled by another code.

FIGURE 5.38

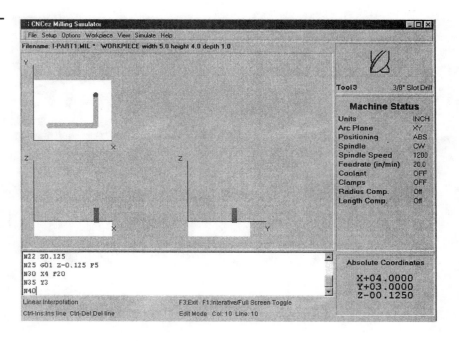

FIGURE 5.39

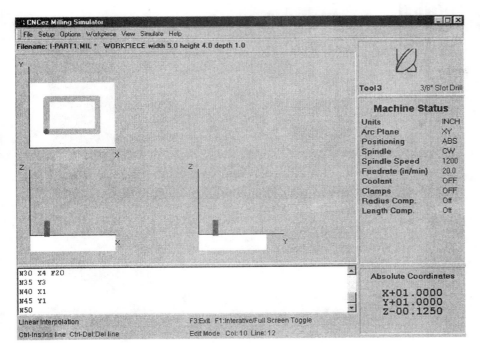

>N45 X1 [Enter]   Feed back 4 in. (Fig. 5.39)

>N50 Y1 [Enter]   Feed down 3 in. (Fig. 5.39)

>N55 G00 Z1 [Enter]   Rapid out to Z

>N60 X0 Y0 [Enter]   Rapid to home position (Fig. 5.40)

FIGURE 5.40

FIGURE 5.41

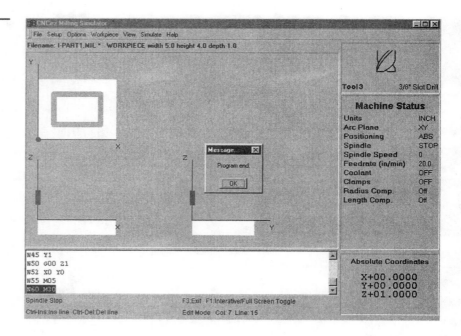

STEP 10: Program shutdown phase. Turn off the spindle and end program (Fig. 5.41).

>N65 M05 [Enter]	Turns spindle off
>N70 M30 [Enter]	End of program
>Press Enter or click on OK	End of Program dialog prompt

STEP 11: The program is now complete. Rerun the program using the Simulate/Cycle option.

From the menu bar, select Simulate, then Cycle Start (Fig. 5.42).

FIGURE 5.42

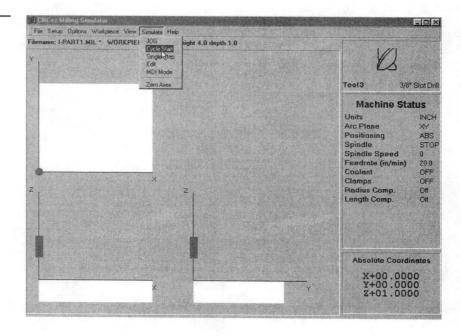

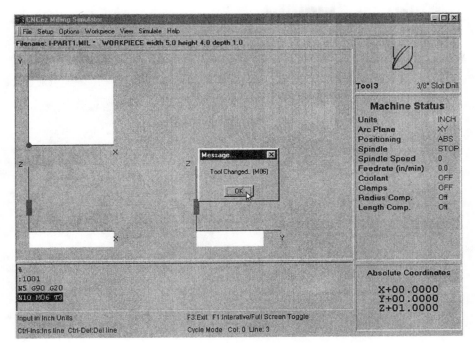

FIGURE 5.43

Note that the program stops at M06 and prompts you for a tool change (Fig. 5.43). Press Enter to continue (Fig. 5.44).

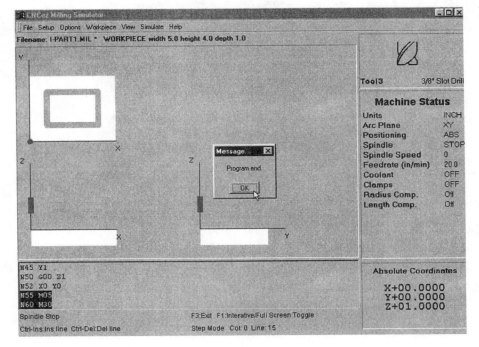

FIGURE 5.44

FIGURE 5.45

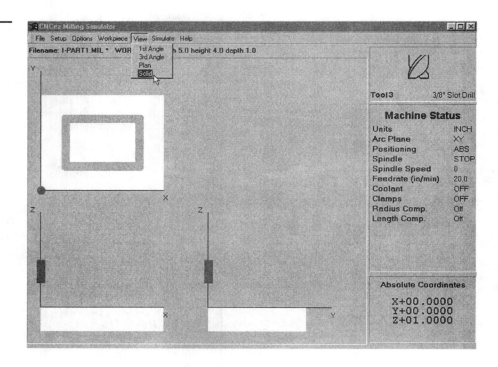

STEP 12: Display the three-dimensional solid, using the View Solid option.

From the pull-down menu, select View.

From View, select Solid (Fig. 5.45).

The final result is shown in Fig. 5.46.

FIGURE 5.46

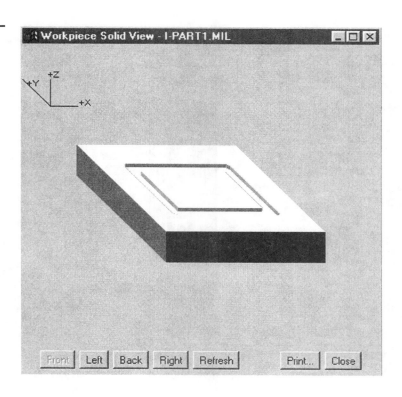

STEP 13: End of program. You can now save the program to a working folder by selecting the File/Save option and creating your own folder, say, "WORKMILL."

EXAMPLE 2: I-part2.mil

This next program introduces you to diagonal linear feed moves, where both the X axis and the Y axis are traversed. See Fig. 5.47.

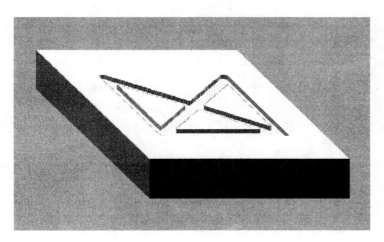

FIGURE 5.47
The finished part. Notice how two axes combined in a G01 feed move results in a diagonal cut. Line 35 is an example of a single-axis feed move, while line 40 is an example of a multiaxis feed move.

Workpiece Size: X5, Y4, Z1

Tool: Tool #2, 1/4" End Mill

Tool Start Position: X0, Y0, Z1 (Relative to workpiece)

```
%
:1002
N5 G90 G20
N10 M06 T2
N15 M03 S1200
N20 G00 X1 Y1
N25 Z0.125
N30 G01 Z-0.125 F5
N35 X4 F10
N40 Y3
N45 X1 Y1
N50 Y3
N55 X4 Y1
N60 G00 Z1
N65 X0 Y0
N70 M05
N75 M30
```

STEP 1: Create a new file called I-part2.
Move the pointer to the menu bar and select File.
Select New.

Type I-part2 and select OK. You are now in the Simulate/ Edit mode.

STEP 2: Set up the workpiece (stock material) for this program.

From the menu bar, select Workpiece.

Select New.

In the Workpiece Definition Dialog enter the workpiece length (in.): 5

Enter the workpiece width (in.): 4

Enter the workpiece height (in.): 1

STEP 3: Program setup phase. You must enter all the setup parameters before you can enter the actual cutting moves.

>% [Enter]	Program start flag
>:1002 [Enter]	Program number
>N5 G90 G20 [Enter]	Absolute coordinates and inch
>N10 M06 T3 [Enter]	Tool change and Tool #3
>N15 M03 S1200 [Enter]	Spindle on CW at 1200 rpm

STEP 4: Material removal phase. Begin cutting the workpiece, using G00 and G01.

>N20 G00 X1 Y1 [Enter]	Rapid to (X1, Y1)
>N25 Z0.125 [Enter]	Rapid to Z0.125
>N30 G01 Z-0.125 F5 [Enter]	Feed move Z−.125", at 5 ipm
>N35 G01 X4 F10 [Enter]	Feed move to X4, at 10 ipm
>N40 Y3 [Enter]	Feed move, at 10 ipm
>N45 X1 Y1 [Enter]	Diagonal feed move to (X1, Y1) (Fig. 5.48)

FIGURE 5.48

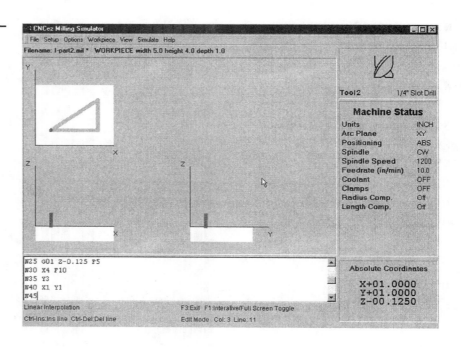

>N50 Y3 [Enter]   Feed move to Y3, at 10 ipm
>N55 X4 Y1 [Enter]   Feed move to (X4, Y1) (Fig. 5.49)
>N60 G00 Z1 [Enter]   Rapid to Z1
>N65 X0 Y0 [Enter]   Rapid to (X0, Y0)

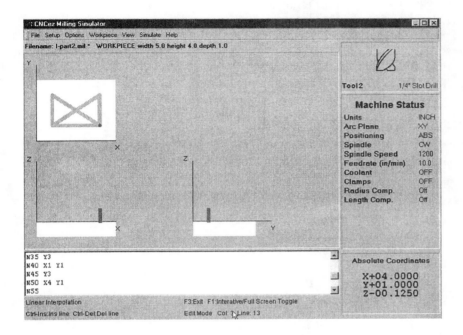

FIGURE 5.49

STEP 5: Program shutdown phase. Turn off the spindle and end program.

>N70 M05 [Enter]   Turn off spindle
>N75 M30 [Enter]   End of program

STEP 6: End of program. You can now optionally save your program as in the previous example.

EXAMPLE 3: I-part3.mil

This program introduces arcs: G02 (clockwise) and G03 (counterclockwise). These are all simple quarter quadrant arcs with a 1-in. radius (see Fig. 5.50).

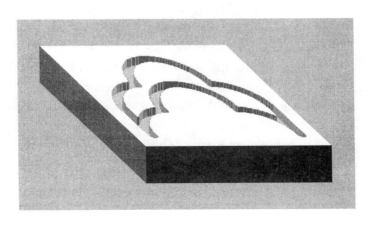

FIGURE 5.50
The completed part.

Workpiece Size: X5, Y4, Z1

Tool: Tool #2, 0.25" Slot Mill

Tool Start Position: X0, Y0, Z1 (Relative to workpiece)

```
%
:1003
N5 G90 G20
N10 M06 T2
N15 M03 S1200
N20 G00 X0.5 Y0.5
N25 Z0.25
N30 G01 Z-0.25 F5
N35 G02 X1.5 Y1.5 I1 J0 F10
N40 X2.5 Y2.5 R1
N45 X3.5 Y1.5 I0 J-1
N50 X4.5 Y0.5 R1
N55 G01 Y1.5
N60 G03 X3.5 Y2.5 R1
N65 X2.5 Y3.5 I-1 J0
N70 X1.5 Y2.5 R1
N75 X0.5 Y1.5 I0 J-1
N80 G01 Y0.5
N85 G00 Z1
N90 X0 Y0
N95 M05
N100 M30
```

STEP 1:   Create a new file called I-part3.

Move the pointer to the menu bar and select File.

Select New.

Enter the file name: I-part3 [Enter] or select OK.

STEP 2:   Set up the workpiece (stock material) for this program.

From the menu bar, select Workpiece.

Select New.

Enter the workpiece length (in.): 5

Enter the workpiece width (in.): 4

Enter the workpiece height (in.): 1

Click on OK

STEP 3:   Change the view to Plan view, using View/Plan

From the menu bar, select View.

Select Plan, as shown in Fig. 5.51.

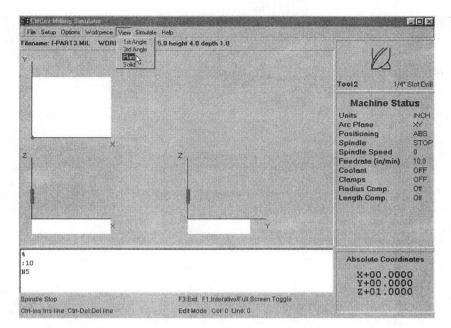

FIGURE 5.51

STEP 4: Program setup phase. You must enter all the setup parameters before you can enter the actual cutting moves.

>% [Enter]	Program start flag
>:1003 [Enter]	Program number
>N5 G90 G20 [Enter]	Absolute coordinates and inch
>N10 M06 T3 [Enter]	Tool change and Tool #3
>N15 M03 S1200 [Enter]	Spindle on CW at 1200 rpm

STEP 5: Material removal phase. Begin cutting the workpiece, using G00, G01, G02, and G03.

>N20 G00 X0.5 Y0.5 [Enter]	Rapid to (X0.5, Y0.5)
>N25 Z0.25 [Enter]	Rapid to Z0.25
>N30 G01 Z-0.25 F5 [Enter]	Feed down to −0.25 at 5 ipm
>N35 G02 X1.5 Y1.5 I1 J0 F10 [Enter]	Circular interpolation (Fig. 5.52)
>N40 X2.5 Y2.5 R1 [Enter]	G02 command, using Radius (Fig. 5.53)
>N45 X3.5 Y1.5 I0 J-1 [Enter]	G02 command, using I and J values
>N50 X4.5 Y0.5 R1 [Enter]	G02 command, using R value
>N55 G01 Y1.5 [Enter]	G01 feed to Y1.5 (Fig. 5.54)
>N60 G03 X3.5 Y2.5 R1 [Enter]	G03 arc, using R value
>N65 X2.5 Y3.5 I-1 J0 [Enter]	G03 arc, using I and J values
>N70 X1.5 Y2.5 R1 [Enter]	G03 arc, using R value
>N75 X0.5 Y1.5 I0 J-1 [Enter]	G03 arc, using I and J values

FIGURE 5.52
G02 Circular interplation clockwise.

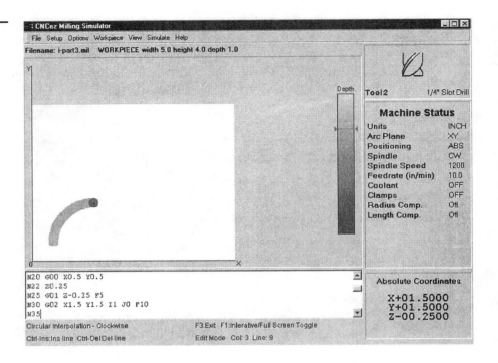

FIGURE 5.53
G02 move using R1 rather than I1, J0.

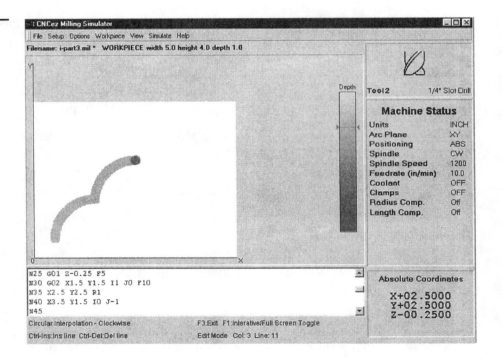

>N80 G01 Y0.5 [Enter]   G01 feed move to Y0.5

>N85 G00 Z1 [Enter]   G00 rapid move to Z1

>N90 X0 Y0 [Enter]   G00 rapid move to X0,Y0 (Fig. 5.55)

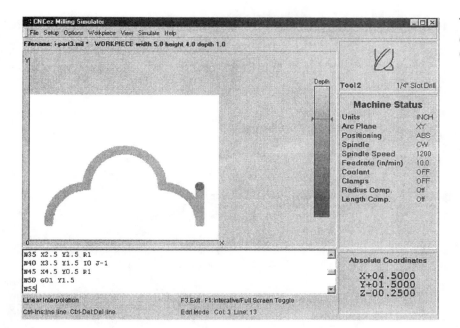

FIGURE 5.54
Completed G02 operation with
G01 move to Y1.5.

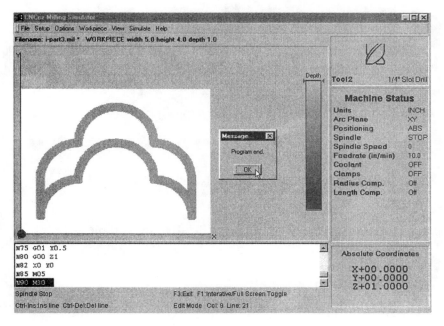

FIGURE 5.55
The completed part.

STEP 6: Program shutdown phase. Turn off the spindle and end program.

>N85 M05 [Enter]   Turn off spindle

>N90 M30 [Enter]   End of program. You can now optionally save your program to your working folder.

STEP 7: End of program

EXAMPLE 4: I-Part4.Mil

This program cuts several G02 and G03 arcs (clockwise and counterclockwise) in semicircles and full circles (see the finished part in Fig. 5.56).

FIGURE 5.56
Solid ISO view of the finished part.

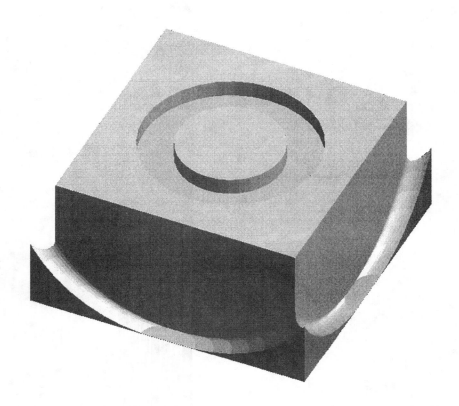

Workpiece Size: X4, Y4, Z2

Tool: Tool #4, 0.5" Slot Mill

Tool Start Position: X0, Y0, Z1 (Relative to workpiece)

```
%
:1004
N5 G90 G20
N10 M06 T4
N15 M03 S1200
N20 G00 Z0.25
N25 G01 Z0 F5
N30 G18 G02 X4 Z0 I2 K0
N35 G19 G03 Y4 Z0 J2 K0
N40 G18 G03 X0 Z0 I-2 K0
N45 G19 G02 Y0 Z0 J-2 K0
N50 G00 Z0.25
N55 X1 Y2
```

(continues)

(continued)

```
N60 G01 Z-0.25
N65 G17 G02 I1 J0 F10
N70 G00 Z1
N75 X0 Y0
N80 M05
N85 M30
```

STEP 1:   Create a new file called I-part4.

Move the pointer to the menu bar and select File.

Select New.

Enter the file name: I-part4 [Enter] or click on OK.

STEP 2:   Set up the workpiece (stock material) for this program and save the workpiece as a stock part.

From the menu bar, select Workpiece.

Select New.

Enter the workpiece length (in.): 4

Enter the workpiece width (in.): 4

Enter the workpiece height (in.): 2

Click on OK

From the menu bar, select Workpiece.

Select Save (Fig. 5.57).

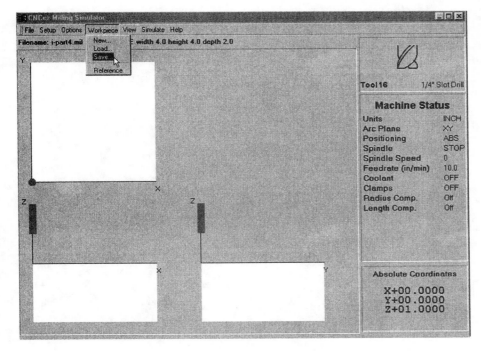

FIGURE 5.57
Selecting Save from the Work-piece menu.

FIGURE 5.58
Entering the workpiece filename.

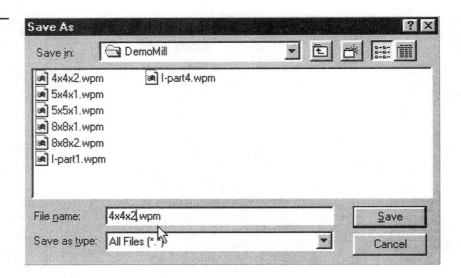

Enter the file name: 4x4x2-in.wpm [Enter] or Click on Save
(Fig. 5.58).

STEP 3:   Make sure Showpaths are on in order to show the centerline
moves.

From the Menu Bar, select Options.

From the Options menu, select Show Toolpaths (Fig. 5.59).

Also, be sure to switch to 3rd Angle View.

From the Menu Bar, select View/3rd Angle.

FIGURE 5.59
Selecting Show Toolpath option.

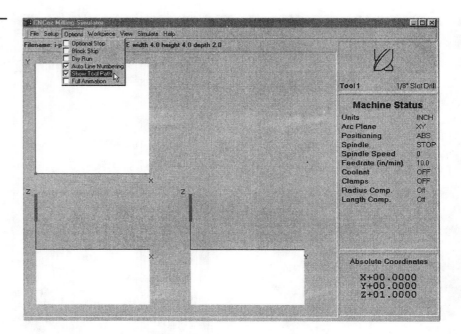

STEP 4: Program setup phase. You must enter all the setup parameters before you can enter the actual cutting moves.

>% [Enter]	Program start flag
>:1004 [Enter]	Program number
>N5 G90 G20 [Enter]	Absolute coordinates and inch
>N10 M06 T4 [Enter]	Tool change and Tool #4
>N15 M03 S1200 [Enter]	Spindle on CW at 1200 rpm

STEP 5: Material removal phase. Begin cutting the workpiece, using G00, G01, G02, and G03.

>N20 G00 Z0.25 [Enter]	Rapid to Z0.25
>N25 G01 Z0 F5 [Enter]	Feed down to Z0 at 5 ipm
>N30 G18 G02 X4 Z0 I2 K0 [Enter]	Circular interpolation, using G18 on the X and Z arc plane (Fig. 5.60)

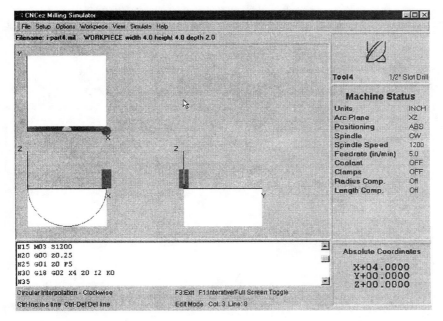

FIGURE 5.60
Part after the G18 and G02 arc move.

>N35 G19 G03 Y4 Z0 J2 K0 [Enter]	Circular interpolation, using G19 in the Y and Z arc plane (Fig. 5.61)
>N40 G18 G03 X0 Z0 I-2 K0 [Enter]	Circular interpolation, using G18 in the X and Z arc plane (Fig. 5.62)

FIGURE 5.61
Part after the G19 and G03 arc
feed move.

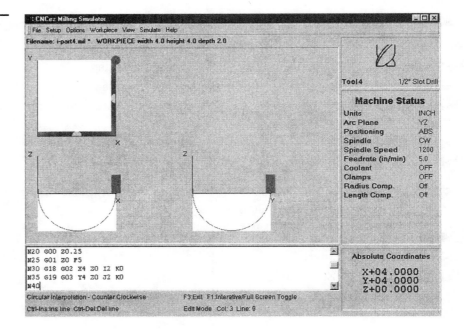

FIGURE 5.62
Part after G19 and G02 arc feed
move.

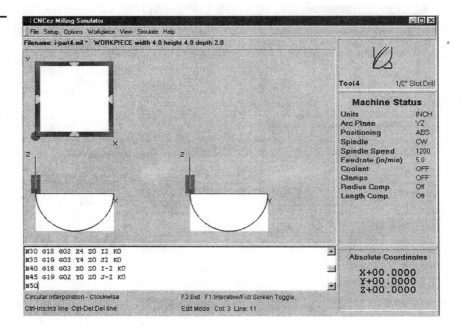

>N45 G19 G02 Y0 Z0 J-2 K0 [Enter]

Circular interpolation, using G19 in the Y and Z arc plane (Fig. 5.62)

>N50 G00 X1 Y2 Z0.25 [Enter]

Rapid up to (X1, Y1, Z0.25)

>N55 X1 Y2

Rapid to (X1, Y2)

>N60 G01 Z-0.25 [Enter]

Feed in to Z−0.25 at 5 ipm

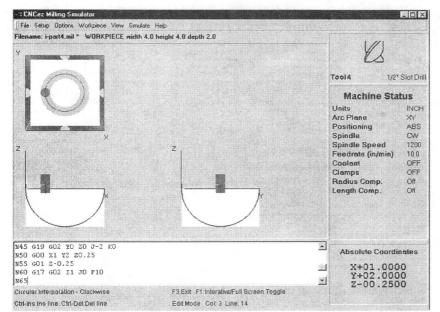

FIGURE 5.63
Part after G17 and G02 arc feed move.

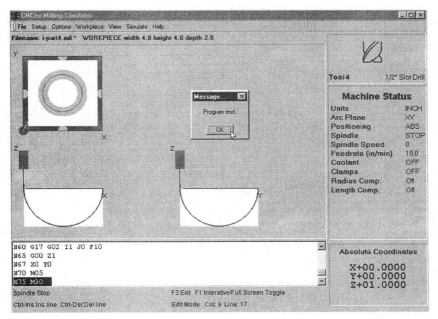

FIGURE 5.64
Completed part and end of program.

>N65 G17 G02 I1 J0 F10 [Enter]   Circular interpolation in the X and Y arc plane, using G17 (Fig. 5.63)

Note that the G17 command is required; otherwise, the G19 would still be in force.

>N70 G00 Z1 [Enter]   Rapid to Z1
>N75 X0 Y0 [Enter]   Rapid to (X0, Y0)

STEP 6: Program shutdown phase. Turn off the spindle and end program.

>N80 M05 [Enter]   Turn off spindle
>N85 M30 [Enter]   End of program (Fig. 5.64)

STEP 7: End of Program. Save Part Program to working folder.

EXAMPLE 5: I-part5.mil

This program involves a simple drilling cycle with a defined retract plane. Once the G-code for the drill cycle has been executed, only the X and/or Y location of the remaining holes need to be defined (see the finished part in Fig. 5.65).

FIGURE 5.65
The finished part.

Workpiece Size:	X5, Y4, Z1
Tool:	Tool #7, 3/8" HSS Drill
Tool Start Position:	X0, Y0, Z1 (Relative to workpiece)

```
%
:1005
N5 G90 G20
N10 M06 T7
N15 M03 S1000
N20 G00 X1 Y1
N25 Z0.25
N30 G98 G81 X1 Y1 Z-0.25 R0.25 F3
N35 Y2
N40 Y3
N45 X2
N50 Y2
N55 Y1
N60 X3
```

(continues)

(continued)

```
N65  X4
N70  Y2
N75  Y3
N80  X3
N85  Y2
N90  G00  Z1
N95  X0  Y0
N100 M05
N105 M30
```

STEP 1: Create a new file called I-part5.

Move the pointer to the menu bar and select File.

Select New.

Enter the file name: I-part5 [Enter] or Click on OK.

STEP 2: Set up the workpiece (stock material) for this program and save the workpiece as a stock part.

From the menu bar, select Workpiece.

Select New.

Enter the workpiece length (in.): 5

Enter the workpiece width (in.): 4

Enter the workpiece height (in.): 1

STEP 3: Be sure to turn Showpaths on in order to show the centerline moves.

From the menu bar, select Options.

From Options, select Show Toolpaths.

STEP 4: Program setup phase. You must enter all the setup parameters before you can enter the actual cutting moves.

>% [Enter]	Program start flag
>:1005 [Enter]	Program number
>N5 G90 G20 [Enter]	Absolute coordinates and inch
>N10 M06 T7 [Enter]	Tool change and Tool #7
>N15 M03 S1000 [Enter]	Spindle on CW at 1000 rpm

STEP 5: Material removal phase

>N20 G00 X1 Y1 [Enter]	Rapid to (X1, Y1)
>N25 Z0.25 [Enter]	Rapid to Z0.25
>N30 G98 G81 X1 Y1 Z-0.25 R0.25 F3 [Enter]	G81 drill cycle with the first hole at (X1, Y1, Z–0.25) and the retract at Z0.25, (Figs. 5.66 and 5.67)
>N35 Y2 [Enter]	The drill moves to Y2, drills in to –0.25, and then retracts to Z0.25 (Fig 5.68)

FIGURE 5.66
The workpiece before the drill cycle.

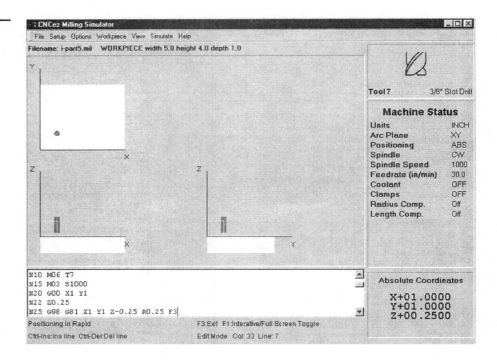

FIGURE 5.67
The workpiece after the drill cycle.

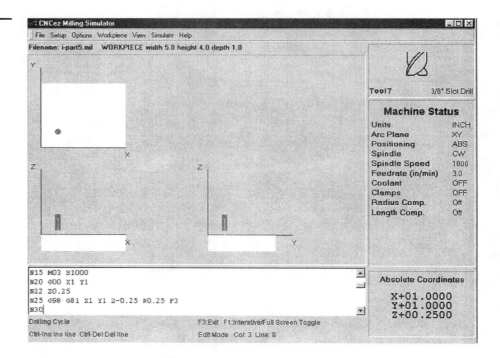

>N40 Y3 [Enter]   Part of drill cycle (Fig. 5.69)

>N45 X2 [Enter]   Fig. 5.69

>N50 Y2 [Enter]   Fig. 5.69

>N55 Y1 [Enter]   Fig. 5.69

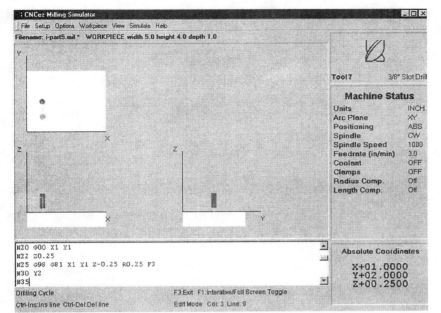

FIGURE 5.68
The second hole by simply
entering Y2.

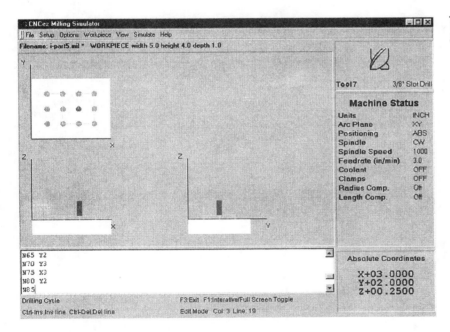

FIGURE 5.69
Shows all holes drilled.

>N60 X3 [Enter]   Fig. 5.69

>N65 X4 [Enter]   Fig. 5.69

>N70 Y2 [Enter]   Fig. 5.69

>N75 Y3 [Enter]   Fig. 5.69

>N80 X3 [Enter]   Fig. 5.69

>N85 Y2 [Enter]   Fig. 5.69

>N90 G00 Z1 [Enter]   Rapid to Z1

>N95 X0 Y0 [Enter]   Rapid to (X0, Y0)

STEP 6: Program shutdown phase. Turn off the spindle and end program.

>N100 M05 [Enter]   Turn off spindle
>N105 M30 [Enter]   End of program (Fig. 5.70)

STEP 7: End of program.   OPTIONALLY   Save your program to your working folder.

FIGURE 5.70
The end of the program.

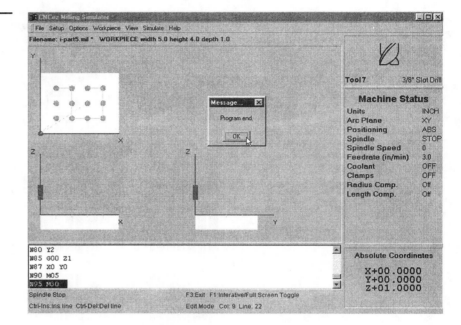

EXAMPLE 6: I-part6.mil

This program involves a drilling cycle with a dwell and incremental coordinates (see the finished part in Fig. 5.71).

FIGURE 5.71
The finished part.

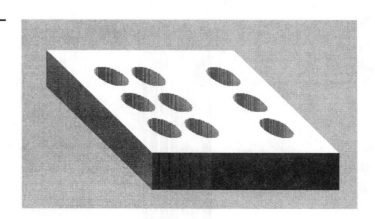

Workpiece Size: X5, Y4, Z1

Tool: Tool #8, 3/4" HSS Drill

Tool Start Position: X0, Y0, Z1 (Relative to workpiece)

```
%
:1006
N5 G90 G20
N10 M06 T8
N15 M03 S500
N20 G00 X1 Y1
N25 Z0.25
N30 G91 G98 G82 Z-0.5 R0.25 P1
N35 X1
N40 X2
N45 Y1
N50 Y1
N55 X-2
N60 X-1
N65 Y-1
N70 X1
N75 G80 G90 G00 Z1
N80 X0 Y0
N85 M05
N90 M30
```

STEP 1:   Create a new file called I-part6.
Move the pointer to the menu bar and select File.
Select New.
Enter the file name: I-part6 [Enter] or Click on OK.

STEP 2:   Load the workpiece (stock material) for this program.
From the menu bar, select Workpiece.
Select Load.
Select the file: 5X4X1-IN (Figs. 5.72 and 5.73).

STEP 3:   Turn Showpaths on in order to show the centerline moves.
From the menu bar, select Options.
From Options, select Show Toolpaths.

STEP 4:   Program setup phase. You must enter all the setup
parameters before you can enter the actual cutting moves.

>% [Enter]	Program start flag
>:1006 [Enter]	Program number
>N5 G90 G20 [Enter]	Absolute coordinates and inch
>N10 M06 T8 [Enter]	Tool change and Tool #8
>N15 M03 S500 [Enter]	Spindle on CW at 500 rpm

FIGURE 5.72

Selecting the stock material work-piece.

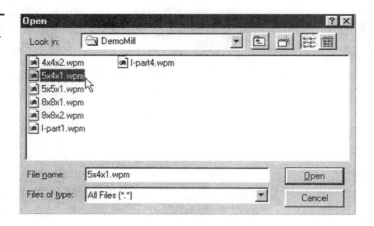

FIGURE 5.73

After selecting the stock workpiece.

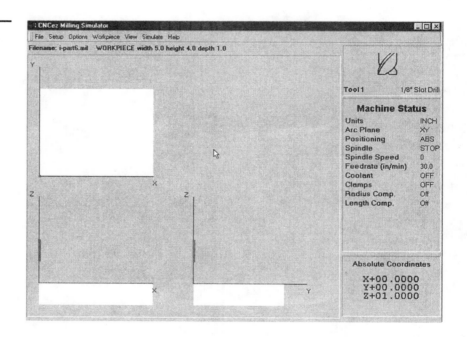

STEP 5:   Material removal phase

>N20 G00 X1 Y1 [Enter]   Rapid to (X1, Y1)

>N25 Z0.25 [Enter]   Rapid to Z0.25 (Fig. 5.74)

>N30 G91 G98 G82 Z-0.5 R0.25 P1 [Enter]   Positioning in incremental mode, the Cutter will drill in to −0.25 in., perform a 1-second dwell, and retract to 0.25 in. above the part (Fig. 5.75)

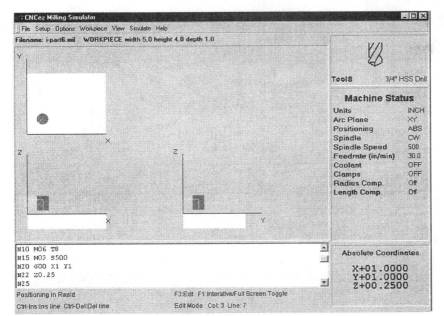

FIGURE 5.74
Rapid to Z0.25.

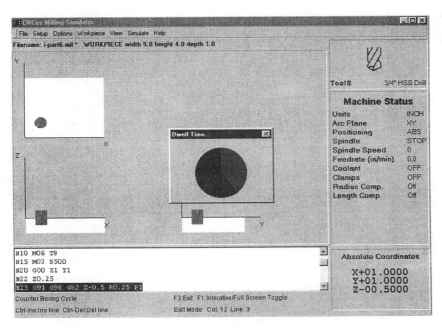

FIGURE 5.75
Result of G98 and G82 commands.

>N35 X1 [Enter]   Drill will move 1 in. to the right, drill in to −0.25 in., perform a 1-second dwell, and retract to 0.25 in. above the part (Fig. 5.76)

>N40 X2 [Enter]   Drill will move 2 in. to the right, drill in to −0.25 in., perform a 1-second dwell, and retract to 0.25 in. above part (Fig. 5.77)

>N45 Y1 [Enter]   Drill will move 1 in. on the Y axis, drill in to −0.25 in., perform a 1-second dwell, and retract to 0.25 in. above the part.

FIGURE 5.76
Entering X1 causes the cutter to move to the right 1 in. and drill down with a dwell and finally return to the retract plane.

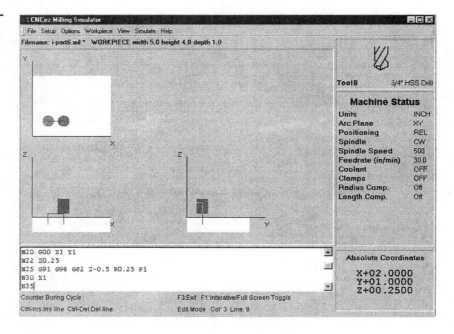

FIGURE 5.77
Result after entering X2 on N35.

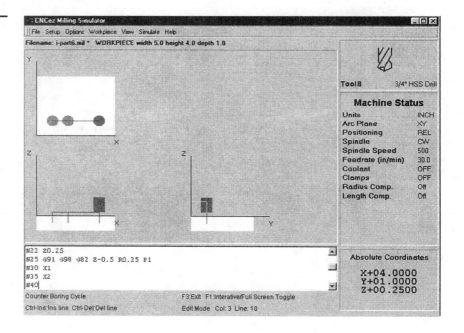

>N50 Y1 [Enter]   Drill will move 1 in. on the Y axis, drill in to −0.25 in., perform a 1-second dwell, and retract to 0.25 in. above the part (Fig. 5.78)

>N55 X-2 [Enter]

>N60 X-1 [Enter]

>N65 Y-1 [Enter]

>N70 X1 [Enter]   (Fig. 5.79)

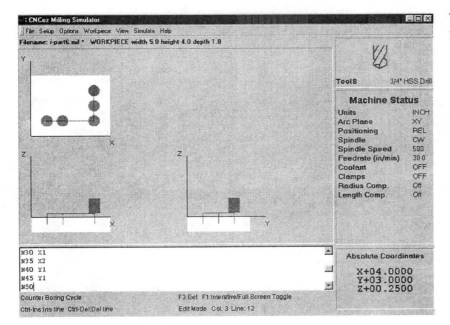

FIGURE 5.78
Result after entering Y1 on N45.

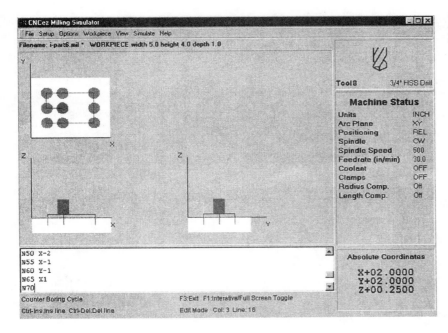

FIGURE 5.79
Result after final drill on N65.

>N75 G80 G90 G00 Z1 [Enter]   Cancels the drill cycle and returns to absolute programming

>N80 X0 Y0 [Enter]   (Fig. 5.80)

STEP 6: Program shutdown phase. Turn off the spindle and end program.

>N85 M05 [Enter]   Turn off spindle

>N90 M30 [Enter]   End of program (Fig. 5.81)

STEP 7: End of program   You can optionally save your part program to your working folder.

FIGURE 5.80
Result after entering the drill
cycle cancel and G90 absolute
with a rapid move.

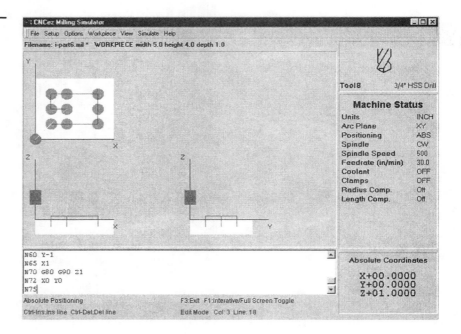

FIGURE 5.81
End of program dialog prompt.

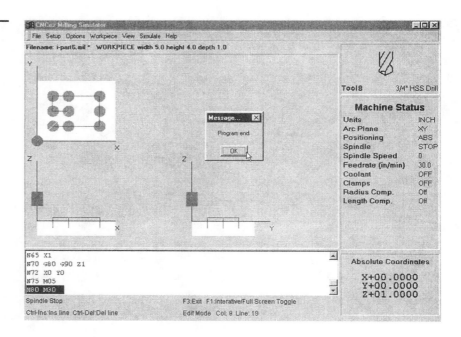

LAB EXERCISES

1. What does the preparatory function G01 command do?

2. How are tool length offsets called?

3. Give an example of a positioning in rapid move.

4. What does the letter address F stand for when a G01 command is programmed?

5. What plane does the G18 command specify?

6. Which G-code and additional letter address call up cutter compensation left?

7. Write an example start line for a G81 drilling cycle.

8. Which M-code specifies program end, reset to start?

CHAPTER 6

Turning

CHAPTER OBJECTIVES

After studying this chapter, the student should have knowledge of the following:

Programming for turning operations

Linear and circular interpolation programming

Tool nose radius compensation

Word address commands for the NC Lathe

Use of Multiple repetitive cycles

When working through this chapter, have the software running on your computer and enter the programs by using the Simulate/Edit or (MDI) options.

Follow these steps when simulating programming codes or creating new programs:

REMEMBER

1. Select New from File menu.
   Enter a file name.
2. Change the tool library if required.
3. Select the Workpiece menu.
   Specify or load a new workpiece.
4. Select the Simulate menu.

Use the Simulate/Edit mode if you want to create a program or run it or the Simulate/MDI mode simply to test a code.

Note: The completed sample programs are in the *Demoturn* folder.

LETTER ADDRESS LISTING

Letter addresses are variables used in G- and M-codes. Most G-codes contain a variable, defined by the programmer, for each specific function. Each letter used in CNC programming is called an address, or word. The words used for programming are as follows.

F	Assigns a feedrate
G	Preparatory function
I	X-axis location of ARC center
K	Z-axis location of ARC center
M	Miscellaneous function
N	Block number (specifies the start of a block)
P	Start block
D	Dwell time
Q	Block end
D	Depth
R	Radius
S	Sets the spindle speed
T	Specifies the tool to be used
A	Tool angle
U	X stock or incremental value
W	Z stock or incremental value
X	X coordinate
Z	Z coordinate

Word addresses are described in more detail next.

Letter	Address	Description
F	Feedrate	Specifies a feedrate in units (inches or millimeters) per minute or per revolution. In threading, designates the thread pitch. Note that some controllers use F to designate the thread lead.
G	Preparatory function	Specifies a preparatory function. Allows for various modes (for example, rapid and feed) to be set during a program. These will be described in the following sections.
I	X value for G02/G03	Specifies the X distance and direction from a startpoint to a centerpoint on the X axis (see G02 and G03).
K	Z value for G02/G03	Specifies the Z distance and direction from a startpoint to a centerpoint on the Z axis (see G02 and G03). Also can specify the final depth of a thread during threading operations.
M	Miscellaneous code	Controls coolant on/off, spindle rotation, and the like. These will be described in the following sections.
N	Block number	Specifies a block, or sequence, number. Used for program line identification. Allows the programmer to organize each line and is helpful while you are editing. In CNCez, increment default of five to allow extra lines to be inserted if needed during editing.
P	Start block	Used within multiple repetitive cycles to specify the block number of the first block of the finish pass definition.
	Dwell time	Can also specify the length of time in seconds in a dwell command (see G04).
Q	Block end	Used within multiple repetitive cycles to specify the block number of the last block of the finish pass definition.
D	Depth	Specifies the depth of cut of the first pass in threading.
R	Radius	For circular interpolation, replaces the I and K to provide an easier way to designate the radius of a circular movement.
S	Spindle speed	Specifies the spindle speed in revolutions per minute or surface units (feet or meters) per minute.
T	Tool number	Specifies the turret library position and the offset register number to which to be indexed. (For example, to index Tool #3 and call offset #3, program T0303.)

A	Tool angle	Used in a threading cycle to specify the tool angle.
U	X stock	Used in multiple repetitive cycles (see G70, G71, and G72) to specify the amount of stock to be left on the face for finishing.
W	Z stock	Used in multiple repetitive cycles (see G70, G71, and G72) to specify the amount of stock to be left on the face for finishing.
X	X axis	Designates a coordinate along the X axis.
Z	Z axis	Designates a coordinate along the Z axis.
/	Block skip	Used to skip blocks of CNC code. If the Block Skip switch is ON and a "/" is encountered at the start of a CNC block line, the whole block line will be ignored. This will be described in more detail in the following sections.
( )	Comments	Any code or text inside the "(" and ")" brackets is treated as a comment. This is useful for operator instruction prompts.

G-CODES

G-codes are preparatory functions, which involve actual tool moves (for example, control of the machine). These include rapid moves, feed moves, radial feed moves, dwells, and roughing and profiling cycles. Most G-codes described here are modal, meaning that they remain active until canceled by another G-code. The following codes are described in detail in the following sections:

G00	Positioning in rapid	Modal
G01	Linear interpolation	Modal
G02	Circular interpolation (CW)	Modal
G03	Circular interpolation (CCW)	Modal
G04	Dwell	
G20	Inch units	Modal
G21	Metric units	Modal
G28	Automatic zero return	
G29	Return from zero return position	
G32	Simple or constant thread	
G40	Tool nose radius compensation cancel	Modal
G41	Tool nose radius compensation left	Modal
G42	Tool nose radius compensation right	Modal
G50	Coordinate system setting	
G54-G59	Workpiece coordinate systems	
G70	Finishing cycle	Modal
G71	Turning cycle	Modal
G72	Facing cycle	Modal
G74	Peck drilling cycle	Modal

G75	Grooving cycle	
G76	Threading cycle	
G90	Absolute programming	Modal
G91	Incremental programming	Modal
G96	Constant surface speed setting	Modal
G97	Constant spindle speed setting	Modal
G98	Linear feedrate per time	Modal
G99	Feedrate per revolution	Modal

G00 POSITIONING IN RAPID

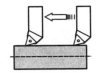

Format: N_ G00 X_ Z_

The G00 command is used primarily to move the tool to and from a noncutting position. It is used most often before and after a G01, G02, and G03 command. It can also be used to position the tool for a tool change.

This command causes the tool to move at its fastest possible rate. This rate may vary among machines, but it will be the fastest possible. On some machines each axis can run independently to exhaust the move as soon as possible. With G00, you can control one or two axes on one block of code. When programming two axes on one line, keep in mind that some controllers will move the tool in a straight line to the endpoint, whereas others will move each motor at its fastest possible rate and thereby cause the motion to be at a 45° angle and then straight. Care must also be taken when you are machining shafts on the outside of the part or performing internal operations. Hence it is advisable to perform single axis rapid moves to avoid collisions between the tool and the workpiece.

REMEMBER

EXAMPLE: N15 G90 G00 X1.8 Z0.5

In this example the tool positions in rapid mode from its present location to a point at X1.8, Z0, as shown in Fig. 6.1.

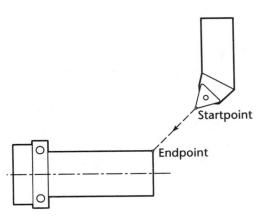

FIGURE 6.1
The G00 causes the tool to move to the endpoint in rapid mode. For safety reasons the G00 should not be used to approach a workpiece.

Sample Program G00EX0:

Workpiece Size: Length 4", Diameter 2"

Tool: Tool #2, Right-hand Facing Tool

Tool Start Position: X2, Z3

`%`	(Program Start Flag)
`:1000`	(Program number 1000)
`N5 G20 G40`	(INCH UNITS, TNR cancel)
`N10 T0101`	(TOOL CHANGE, Tool #1, Register #1)
`N15 M03`	(Spindle on CW)
`N20 G00 X1.9`	(Rapid to X1.9)
`N25 Z0.1 M08`	(Rapid to Z0.1)
`N30 G01 Z-1.25 F0.012`	(Feed to Z-1.25 at 0.12 IN./rev)
`N35 G00 X4.0 Z3.0 M09`	(Rapid to X4 and Z3)
`N40 T0100 M05`	(Tool cancel Register #1)
`N45 M30`	(Program end)

Note: For safety and good programming practice reasons, you should never let the tool be moved rapidly to the workpiece.

G01 LINEAR INTERPOLATION

Format: `N_ G01 X_ Z_ F_`

The G01 command executes all movement along a straight line at a particular feedrate. These straight-line feed moves may cut in one or two axes simultaneously. This command can be used for turning, facing, and tapering.

It is specified by the G01 command, followed by the endpoint of the move, and then a specified feedrate.

EXAMPLE: `G01 Z-2.5 F0.01`

In this example, the tool cuts a straight line from its present location to a point at Z−2.5 at a rate of 0.01 ipr (see Fig. 6.2).

FIGURE 6.2
Normal cutting is performed by using controlled feed moves such as the G01 command.

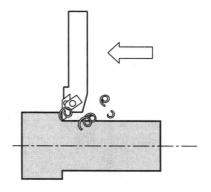

Sample Program G01EX1:

Workpiece Size: Length 4", Diameter 2.5"

Tool: Tool #1, Right-hand Facing Tool

Tool Start Position: X2, Z3

```
%
:1001
N5 G20 G40
N10 T0101
N15 M03
N20 G00 X2.375 M08
N22 Z0.1
N25 G01 Z-2.0 F0.015    (Feed to Z–2.0 at a feedrate of 0.015 ipr)
N30 G00 X2.5
N35 Z0.1
N40 X2.25
N45 G01 Z-1.75          (Feed to Z–1.75 at same feedrate)
N50 G00 X2.375
N55 Z0.1
N60 X2.125
N65 G01 Z-1.5           (Feed to Z–1.5 at same feedrate)
N70 G00 X2.25
N75 Z0.1
N80 X1.875
N85 G01 Z0              (Feed to 0 at same feedrate)
N90 X2.125 Z-0.125      (Feed to (X2.125, Z–1.125) at same
                         feedrate)
N95 G00 X4 M09
N100 Z3
N105 T0100 M05
N110 M30
```

G02 CIRCULAR INTERPOLATION (CW)

Format: N_ G02 X_ Z_ I_ K_ F_

or   N_ G02 X_ Z_ R_ F_

The G02 command executes all circular or radial cuts in a clockwise motion. It is specified by the G02 command, followed by the endpoint for the move, the radius (the distance from startpoint to the centerpoint), and a feedrate. Therefore the three requirements for cutting arcs are:

1. The endpoint.
2. The radius R or I for X and K for Z values that represent the incremental distance from the startpoint to the centerpoint. The R value is limited to a maximum of 90°.
3. The feedrate.

REMEMBER

The I and K values represent the relative, or incremental, distance from the starting point to the arc center.

FIGURE 6.3
Example of the G02 command.
The Startpoint, Centerpoint, and
Endpoint are shown.

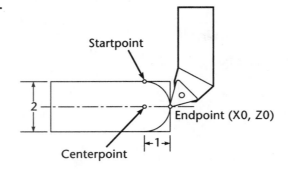

The radius is specified by defining the incremental distance from the arc's startpoint to the centerpoint, in both the X and Z directions. These values are identified by I and K variables, respectively. The R word, the value of the radius of the arc, can also be used. However, the maximum arc is limited to 90° when you use the R word.

EXAMPLE: N05 G01 X2 Z-1 F0.012

                   N10 G02 X0 Z0 I-1 K0

or   N10 G02 X0 Z0 R0.5

In this example, the tool cuts a clockwise arc from its present location (X2, Z-1) at a feedrate of 0.012 ipr (see Fig. 6.3). The tool cuts an arc from (X2, Z–1) to the specified endpoint at (X0, Z0). There is a change of 1 in. in the X direction from the arc's startpoint to its centerpoint, so the I value is –1. There is no change in the Z direction from the startpoint to the centerpoint, so the K value is 0.

Sample Program G02EX2:

Workpiece Size: Length 4", Diameter 2"

Tool: Tool #2, Right-hand Facing Tool

Tool Start Position: X2, Z3

```
%
:1002
N5 G20 G40
N10 T0202
N15 M03
N20 G00 X1.7
N22 Z0.1 M08
N25 G01 Z-0.5 F0.012
N30 G00 X2
N35 Z0.1
N40 X1.4
N45 G01 Z-0.25
N50 G00 X2.1
N55 Z-1
N60 G01 X2.0
N65 G02 X0 Z0 I-1.0 K0        (90° CW arc feed move)
N70 G00 X2.1
```

```
N75  Z-1.0
N80  G01 X2.0
N85  G02 X2.0 Z-2 I0.5 K-0.5  (Partial CW arc feed move)
N90  G00 X4.0 Z3.0 M09
N95  T0200 M05
N100 M30
```

G03 CIRCULAR INTERPOLATION (CCW)

Format: N_ G03 X_ Z_ I_ K_ F_

or   N_ G03 X_ Z_ R_ F_

The G03 command executes all radial cuts in a counterclockwise motion. It is specified by the G03 command, followed by the endpoint for the move, the radius (the distance from the startpoint to the centerpoint), and a feedrate.

The radius is specified by defining the incremental distance from the arc's startpoint to its centerpoint in both the X and Z directions. These values are identified by I and K variables, respectively. The R word, the value of the radius of the arc, can also be used.

EXAMPLE: N10 G01 X0 Z0 F0.012
               N15 G03 X2.0 Z-1.0 I0 K-1.0

In this example, the tool cuts a counterclockwise arc from its present position to (X2, Z–1) at a feedrate of 0.012 ipr (see Fig. 6.4). There is no change in the X direction from the startpoint to the centerpoint, so the I value is 0. There is a –1 difference between the arc's startpoint and its centerpoint in the Z direction, so the K value is –1.

REMEMBER

The I and K values define the incremental distance from the arc startpoint to the arc centerpoint in

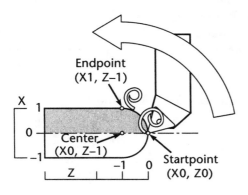

FIGURE 6.4
The tool cuts an arc from (X0, Z0) to the diametrical point (X2, Z-1).

Sample Program G03EX3:

Workpiece Size: Length 4", Diameter 2"

Tool: Tool #2, Right-hand Facing Tool

Tool Start Position: X2, Z3

```
%
:1003
N5 G20 G40
```

(continues)

(continued)

```
N10 T0202
N15 M03
N20 G00 X1.7
N22 Z0.1 M08
N25 G01 Z-0.5 F0.012
N30 G00 X2.0
N35 Z0.1
N40 X1.4
N45 G01 Z-0.25
N50 G00 X1.5
N55 Z0.1
N60 X0
N65 G01 Z0
```
N70 G03 X2.0 Z-1.0 I0 K-1.0   (90° CCW arc feed move)
```
N75 G01 Z-2.0
```
N80 G03 X2.0 Z-1.0 I0.5 K0.5 (Partial CCW arc feed move)
```
N85 G00 X4.0 Z3.0 M09
N90 T0200 M05
N95 M30
```

G04 DWELL

Format: N_ G04 P_

The G04 command designates a period of time that the tool is to wait in a particular position. It is used most frequently in drilling operations.

It is specified by the G04 command, followed by the letter address P, which defines the period of time in seconds.

Upon encountering a G04 command, the MCU halts all axes movement. The spindle, coolant, and other auxiliaries will continue to operate.

EXAMPLE: N10 G04 P2.0

In this example, the tool pauses at its current location for a period of 2 seconds (Fig. 6.5).

FIGURE 6.5
As only the tool motion is affected by the dwell, the spindle will continue to rotate.

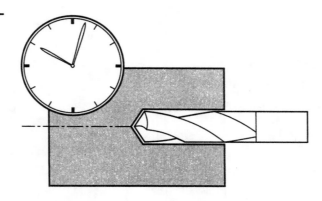

Sample Program G04EX4:

Workpiece Size: Length 4", Diameter 2"

Tool: Tool #8, 1" Diameter Drill

Tool Start Position: X2,Z3

```
%
:1004
N5 G20 G40
N10 T0808
N15 M03
N20 G00 X0
N25 Z0.125
N30 G01 Z-2.0 F0.015
N35 G04 P0.5            (Dwell for 0.5 seconds)
N40 G00 Z3.0
N45 X4.0
N50 T0800 M05
N55 M30
```

G20 INCH UNITS

Format: N_ G20

The G20 command sets the system to accept all data in inch units.

EXAMPLE: N15 G20

In this example, the MCU accepts any and all data as standard inch units. All decimals are indicated by a decimal point, and trailing or leading zeros are optional.

This command may be placed on the same line as other system setting commands (for example, G90 and G40).

This controller rounds all figures to 0.0001 in.

Sample Program G20EX20:

Workpiece Size: Length 4", Diameter 1"

Tool: Tool #1, Right-hand Facing Tool

Tool Start Position: X2, Z3

```
%
:1020
N5 G90
N10 G20                (All data in inch units)
N15 T0101
N20 M03
N25 G00 Z0.1
N30 X0.5
N35 G01 Z0 F0.015
```

(continues)

(continued)

```
N40  G03  X1.0  Z-0.25  IO  K-0.25
N45  G00  X4.0  Z3.0
N50  T0100  M05
N55  M30
```

Block N10 will switch the system to accept all data as inch units. As a result, block N40 will cut a 0.25-in. arc in the workpiece.

G21 METRIC UNITS

Format: N_ G21

The G21 command sets the system to accept all data in metric units, with the millimeter as the standard unit of measure.

EXAMPLE: N35 G21

In this example the MCU accepts any and all data as metric units. This controller requires that all decimals be indicated with a decimal point. Trailing or leading zeros are optional. This command can be placed on the same line as other system-setting commands (for example, G90 and G40).

Note: When working in Metric data units, be sure to use the Metric Tool Library.

Sample Program G21EX21:

Workpiece Size: Length 100 mm, Diameter 50 mm

Tool: Tool #1, Neutral Tool

Tool Start Position: X50, Z75

```
%
:1013
N5   G90  G21  G40              (All data in millimeters)
N15  T0303
N17  M03
N20  G00  Z20  M08
N25  X45
N30  G01  Z-50  F1
N35  X40
N40  Z2
N45  X35
N50  Z-25
N55  X30
N60  Z-10
N65  G00  X100  M09
N70  Z100
N75  T0300  M05
N80  M30
```

G28 AUTOMATIC ZERO RETURN

Format: N_ G28

or   N_ G28 X_ Z_

The G28 command is used for automatic tool changing. Basically, it allows the existing tool to be positioned automatically to the predefined zero return position via an intermediate position.

All axes are positioned to the intermediate point at the rapid traverse rate and then from the intermediate point to the zero return position (Fig. 6.6).

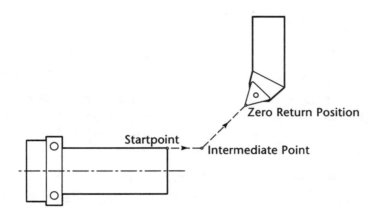

FIGURE 6.6
Cutter moves from the startpoint to the intermediate point and then to the zero return position.

Sample Program G28EX28:

Workpiece Size: Length 4", Diameter 2.5"

Tool: Tool #1, Right-hand Tool

Tool Start Position: X2", Z3"

Zero Return Position: X2", Z3"

```
%
:1028
N5 G20 G40
N10 T0101
N12 M03 M08
N15 G00 X2.25
N20 Z0.1
N25 G01 Z-2.0 F0.012
N30 G28 X4.0 Z0.5 M09 (Rapid move to zero return position)
N35 T0100 M05
N40 M30
```

G29 RETURN FROM ZERO RETURN POSITION

Format: N_ G29

or   N_ G29 X_ Z_

FIGURE 6.7
Cutter moves from the zero reference point to the intermediate point, and then to the endpoint.

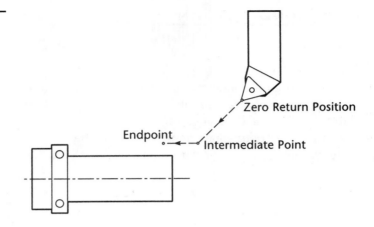

The G29 command is used immediately after a G28 command to return the tool to a specified point via the intermediate point specified by the previous G28 command (Fig. 6.7). This command is usually used after a tool change command.

Sample Program G29EX29:

Workpiece Size:	Length 4", Diameter 2.5"
Tools:	Tool #1, Right-hand Tool
	Tool #2, Finishing Tool
Tool Start Position:	X2", Z3"
Zero Return Position:	X2", Z3"

```
%
:1029
N5  G20
N10 T0101
N12 M03 M08
N15 G00 X2.25
N20 Z0.1
N25 G01 Z-2.0 F0.012
N30 G28 X4.0 Z0.5 M09
N35 T0100 M05
N40 T0202
N45 M03 M08
N50 G29 X2.25 Z0.1      (Return from zero return position)
N55 G01 Z-2.0
N60 G28 X4 Z0.5 M09
N65 T0100 M05
N70 M30
```

G32 SIMPLE THREAD CYCLE

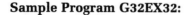

Format: N_ G32 X_ Z_ F_

The G32 command is used to program a simple or basic thread cycle. The X and Z specify the endpoint, F the thread pitch.

Sample Program G32EX32:

Workpiece Size: Length 4", Diameter 2"

Tool: Tool #4, Neutral Tool / Threading Tool

Tool Start Position: X2, Z3

```
%
:1040
N5 G20 G40                  (Set TNR cancel at beginning)
N10 T0101
N15 M03 M08
N20 G00 Z0.1
N25 X2.1
N30 G01 X1.0 F0.012
N35 G32 Z-2.0 F.05          (Machine a single-pass thread)
N40 G00 X4.0
N45 T0100 M05
N50 M30
```

G40 TOOL NOSE RADIUS (TNR) COMPENSATION CANCEL

Format: N_ G40

The G40 command cancels any compensation that was applied to the tool during a program. It also acts as a safeguard to cancel any cutter compensation applied previously.

TNR compensation is used to compensate for the small radius on single-point turning and boring tools. Normally, CNC programs are written so that the tool tip follows the toolpath. This would be fine if all tools were perfect and had sharp points. However, most turning tools have a small radius on the cutting edge that makes the tools deviate slightly from the programmed path. In this case, the tool must be offset either left or right. However, tool nose radius compensation is modal, so the compensation must be canceled once it is no longer required. The G40 command does this.

Sample Program G40EX40:

Workpiece Size: Length 4", Diameter 2"

Tool: Tool #1, Right-hand Facing Tool

Tool Start Position: X2, Z3

(continues)

(continued)

```
%
:1040
N5  G20 G40              (Set TNR cancel at beginning)
N10 T0101
N15 M03 M08
N17 G41
N20 G00 Z-0.1
N25 X2.1
N30 G01 X1 F0.012
N35 G40 G00 Z3 M09      (TNR cancel with rapid to start)
N40 X4
N45 T0100 M05
N50 M30
```

G41 TOOL NOSE RADIUS COMPENSATION LEFT

Format: N_ G41

The G41 command applies a tool nose radius compensation left of the programmed tool path. This allows you to program turning and boring tools by using actual coordinates and not having to allow for the radius of the tool. The offset register value will compensate for the difference caused by the small radius on the tool (Fig. 6.8).

TNR compensation is used to compensate for the small radius on single-point turning and boring tools. The actual offset value on most turning centers comes from the Tool Offset Register Table, which is referenced to the tool currently being used. For example, T0202 would mean that Tool #2 is being used with the offset register #2 loaded. Hence TNR compensation is the value of offset register #2.

Once compensation has been invoked, it remains in effect until canceled by a G40 command or another offset reference. You must remember to cancel tool nose radius compensation when you return a tool to the tool change position, usually after a G28 command.

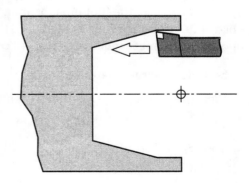

FIGURE 6.8
The tool follows the programmed tool path by automatically compensating the TNR distance on the left side.

Sample Program G41EX41:

Workpiece Size: Length 4", Diameter 2"

Tools: Tool #8, Left-hand Facing Tool

                            Tool #10, 1" Drill

                            Tool #12, Center Drill

Tool Start Position: X2, Z3

```
%
:1041
N5 G20 G40
N10 T1212 M08
N15 G98 M03 S2000
N20 G00 X0
N25 Z0.1
N30 G01 Z-0.2 F2
N35 G00 Z3
N40 X4 T1200
N45 T1010
N50 G00 X0
N55 Z0.1
N60 G74 Z-2 F0.5 D0 K0.25
N65 G00 Z3
N70 T1000 X4
N75 G99 T0808
N80 G41 G00 X0.8          (TNR compensation left)
N85 Z0.2
N90 G01 Z-1.75 F0.012
N95 X0.9
N100 Z-0.25
N105 G03 X2 Z0 R0.5
N110 G40 G00 X4 Z3 M09
N115 T0800 M05
N120 M30
```

G42 TOOL NOSE RADIUS COMPENSATION RIGHT

Format: N_ G42

The **G42** command applies tool nose radius compensation right to allow you to program turning and boring tools by using actual coordinates and not having to allow for the radius of the tool. The offset register value will compensate for the difference caused by the small radius on the tool (Fig. 6.9).

FIGURE 6.9

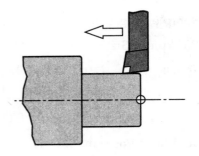

TNR compensation is used to compensate for the small radius on single-point turning and boring tools. The actual offset value on most turning centers comes from the Tool Offset Register Table, which is referenced to the tool currently being used. For example, T0404 would mean that Tool #4 is being used with the offset register #4 loaded. Hence TNR compensation is the value of offset register #4.

Once compensation has been invoked, it remains in effect until canceled by a G40 command. You must remember to cancel tool nose radius compensation when you return a tool to the tool change position.

Sample Program G42EX42:

Workpiece Size: Length 4", Diameter 2"

Tools: Tool #1, Right-hand Facing Tool

                    Tool #2, Right-hand Finishing Tool

Tool Start Position: X2, Z3

```
%
:1042
N5 G20 G40
N10 T0101
N15 M03
N20 G00 Z0.1 M08
N25 X1.75
N30 G01 Z-2 F0.012
N35 G00 X2 Z0.1
N40 X1.5
N45 G01 Z-1.5
N50 G00 X1.75 Z0.1
N55 X1
N60 G01 Z0
N65 G03 X1.5 Z-0.25 I0 K-0.25
N70 G00 X4 Z3 T0100
N75 G00 T0202
N80 G42 G00 X0            (TNR compensation right)
N85 Z0.1
N90 G01 Z0
```

(continues)

(continued)

```
N95  X1
N100 G03 X1.5 Z-0.25 R0.25
N105 G01 Z-1.5
N110 X1.75
N115 Z-2
N120 X2.1
N125 G40 G00 X4 Z3 M09
N130 T0200 M05
N135 M30
```

G50 COORDINATE SYSTEM SETTING OR PROGRAMMING ABSOLUTE ZERO

Format: N_ G50 X_ Z_ S_

The G50 command is used to specify the coordinate system setting for the current program. It specifically determines the coordinate distances in both the X and Z directions from the current tool location to the workpiece origin point. This origin point is also referred to as the program absolute zero and can be specified in several ways. Normally, you would touch-off the workpiece manually by jogging the tool to the edge of the workpiece. At this point you would determine where the origin will be. Usually, the simplest method is to make the right-hand centerpoint of the workpiece to be (X0, Z0). In other cases the chuck face or the machine spindle plate will be used as the absolute program zero. The G50 command also has an optional maximum spindle setting for the current operation, which can be programmed by specifying the S address word.

Sample Program G50EX50:

Workpiece Size: Length 4", Diameter 2"

Tool: Tool #1, Right-hand Facing Tool

Tool Start Position: X2, Z3

```
%
:1040
N5  G20 G40              (Set TNR cancel at beginning)
N10 G50 X2 Z6            (Set Origin after touch-off)
N15 T0101
N16 M03 M08
N18 G41
N20 G00 Z-0.1
N25 X2.1
N30 G01 X1 F0.012
N35 G40 G00 Z3 M09      (TNR cancel with rapid to start)
N40 X4
N45 T0100 M05
N50 M30
```

G54–G59 WORK COORDINATE SYSTEMS

Format: N_ G54 – G59

The G54–G59 commands are used to establish one of six preprogrammed work coordinate systems. These settings reside in special parameter registers in the controller or MCU. Each register has separate X and Z coordinate settings. The use of these commands can be thought of as special G92 commands for specific work areas. They are frequently used when multiple part fixtures are used in a job, where each register can refer to a specific work area.

EXAMPLE: N155 G56

In the example above the MCU sets the third working coordinate system from the G56 register values.

Sample Program G54EX54:

Workpiece Size: Length 4", Diameter 2"

Tools: Tool #1, Neutral Tool

Tool Start Position: X2, Z3

Coordinate System

Register #1:   X0.0 Z-2.0

```
%
:1054
N5 G90 G20
N10 T0101
N15 M03
N20 G00 X2 Z0.1 M08
N25 G54                    (Set the working coordinates)
N80 G00 X4.0 Z3.0 M09
N85 T0200 M05
N90 M30
```

G70 FINISHING CYCLE

Format: N_ G70 P_ Q_
P   Start block of segment
Q   End block of segment
F   Feedrate

The G70 command calculates the finished part profile and then executes a finishing (or profile) pass on the workpiece.

You specify it by entering G70, followed by the letter address P (start block) and the letter address Q (end block), which specify the desired finished profile. Thus you define the finish contour of the part. This command is generally used to finish the part after a G71 or G72 command, which have left allowance for the finishing cycle.

When the G70 command is executed, it reads all program blocks and then formulates a profiling cycle. The finishing pass cut follows the finish contour of the part. At the end of the finishing pass the tool is positioned back to the startpoint and the next block is read and executed.

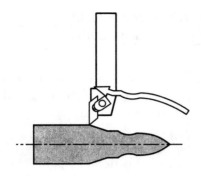

FIGURE 6.10
The tool follows the profile of the part specified by the G70 command. All data between the start and end blocks is considered.

This command is mostly used to profile a part once all material removing operations have been completed.

EXAMPLE: N10 G70 P20 Q45

In this example the MCU executes all commands between N20 and N45, essentially resulting in a profile pass (Fig. 6.10).

Sample Program G70EX70:

Workpiece Size: Length 4", Diameter 2"

Tools: Tool #1, Right-hand Facing Tool

Tool #2, Right-hand Finishing Tool

Tool Start Position: X2, Z3

```
%
:1070
N5 G90 G20
N10 T0101
N15 M03
N20 G00 X2 Z0.1 M08
N35 G71 P40 Q55 U0.05 W0.05 D625 F0.012
N40 G01 X0 Z0
N45 G03 X0.5 Z-0.25 I0 K-0.25
N50 G01 X1.75 Z-1.0
N55 X2.1
N60 T0100 G00 X4.0 Z3.0
N65 T0202
N70 G00 X2 Z0.1
N75 G70 P40 Q55 F0.006        (Finish cycle from N40 to N55)
N80 G00 X4.0 Z3.0 M09
N85 T0200 M05
N90 M30
```

G71 ROUGH TURNING CYCLE

Format: N_ G71_ P_ Q_ U_ W_ D_ F

P	Start block of segment
Q	End block of segment
U	Amount of stock to be left for finishing in X

W   Amount of stock to be left for finishing in Z

D   Depth of cut for each pass in thousandths

F   Feedrate for finish pass

The G71 command automatically turns a workpiece to a specified diameter at a specified depth of cut. It reads a program segment and determines the number of passes, the depth of cut for each pass, and the number of repeat passes for the cycle.

The G71 command can cut four types of patterns. Each must be programmed as a separate group of blocks. Cutting is done by parallel moves of the tool in the Z direction. The U and W signs are shown in Fig. 6.11.

The G71 command must either be inserted into the program after the program segment has been completed or be written in the File Editor. If the command is written in the Simulate/Edit mode, the program will look for the start and end blocks, which do not yet exist, thereby creating errors.

EXAMPLE: N35 G71 P40 Q100 U.025 W.025 D625 F0.012

In this example, the MCU reads the data between N40 and N100 and removes material according to this profile. The cycle has a depth of cut of 1/16 in. for each pass at 0.012 ipr (Fig. 6.12).

FIGURE 6.11
Positive and negative U and W signs instruct the MCU on how to execute the rough turning cycle. This type of cycle is known as a Type A roughing cycle on most Fanuc controllers.

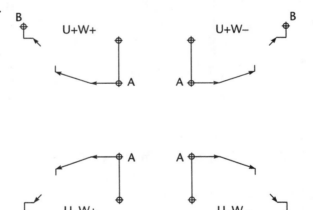

FIGURE 6.12
The G71 command turns the part to the diameter specified in the program segment. The cutting moves are parallel to the Z axis. The upper half of the profile shows the G71 command and how it "steps" down to size. The lower half of the profile shows how the G70 command cleans up with a finishing pass.

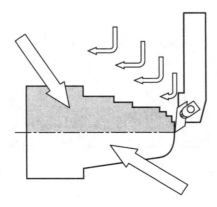

REMEMBER

In the following sample program use the Simulate/Edit option.

Sample Program G71EX71:

Workpiece Size:	Length 4", Diameter 2"
Tools:	Tool #1, Right-hand Facing Tool
	Tool #2, Right-hand Finishing Tool

Tool Start Position: X2, Z3

```
%
:1071
N5 G90 G20
N10 T0101
N15 M03
N20 G00 X2 Z0.1 M08
N35 G71 P40 Q55 U0.05 W0.05 D625 F0.012 (Rough turning)
N40 G01 X0 Z0
N45 G03 X1 Z-0.5 I0 K-0.5
N50 G01 Z-1.0
N55 X2.1 Z-1.5
N60 T0100 G00 X4 Z3
N65 T0202
N70 G00 X2 Z0.1
N75 G70 P40 Q55 F0.006
N80 G00 X4 Z3 M09
N85 T0200 M05
N90 M30
```

G72 ROUGH FACING CYCLE

Format:	N_ G72 P_ Q_ U_ W_ D_ F_
P	Start block
Q	End block
U	Amount of stock to be left for finishing in X
W	Amount of stock to be left for finishing in Z
D	Depth of cut for finish pass
F	Feedrate (this is optional)

The G72 command automatically faces off a part to a predefined depth of cut, with preset offsets and feedrates.

EXAMPLE: N20 G72 P25 Q45 U0.05 W0.05 D500 F0.012

In this example, the MCU reads the commands in N25–N45 and then executes a facing operation, in accordance with the profile defined (Fig. 6.13).

The G72 command must either be inserted into the program after the program segment has been completed or be written in the File Editor. If the

REMEMBER

FIGURE 6.13

The G72 command roughs out the profile as specified in the program segment. The cutting moves are parallel to the X axis. Then the G70 command cleans it up with a finish pass.

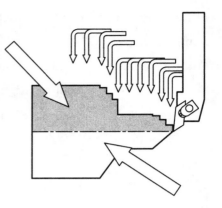

command is written in the Simulate/Edit mode, the program will look for the start and end blocks, which do not yet exist, thereby creating errors.

Sample Program G72EX72:

Workpiece Size:	Length 4", Diameter 1"
Tools:	Tool #1, Right-hand Facing Tool
	Tool #2, Right-hand Finishing Tool

Tool Start Position: X2, Z3

```
%
:1072
N5 G90 G20
N10 T0101
N15 M03
N20 G00 X1 Z0.1 M08
N25 G72 P30 Q50 U0.05 W0.05 D500 F0.012  (Rough facing)
N30 G01 X0 Z0.1
N35 X0.25
N40 Z-0.125
N45 X0.5 Z-0.25
N50 G02 X1 Z-0.5 I0.25 K0
N55 G00 X4 Z3 T0100
N60 T0202
N65 G00 X1 Z0.1
N70 G70 P30 Q50 F0.006
N75 G00 X4
N80 Z3 M09
N85 T0200 M05
N90 M30
```

G74 PECK DRILLING CYCLE

Format: N_ G74 X0 Z_ K_ F_

The G74 command executes a peck drilling cycle with automatic retracts and incremental depths of cut. The G74 command is specified by several letter addresses:

X0	X always 0
Z	Total depth
K	Peck depth
F	Feed rate

EXAMPLE: `N20 G74 X0 Z-1.0 K0.125 F0.015`

In this example, a hole is peck drilled to a total depth of 1 in., using 0.125 in. for the depth of each peck (Fig. 6.14).

Sample Program G74EX74:

Workpiece Size: Length 4", Diameter 1"

Tools: Tool #12, Center Drill

                            Tool #10, 1/2" Drill

Tool Start Position: X2, Z3

```
%
:1074
N5 G20 G98
N10 T1212 M08
N15 M03 S2000
N20 G00 X0
N25 Z0.1
N30 G01 Z-0.2 F2
N35 G00 Z3
N40 X3 T1200
N45 T1010
N50 G00 X0
N55 Z0.1
N60 G74 X0 Z-1.5 K0.125 F0.5          (Drilling cycle)
N65 G00 Z3
N70 X4 M09
N75 T1000 M05
N80 M30
```

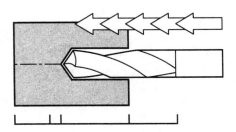

FIGURE 6.14
The G74 command relies on the command variables to execute properly. Z is the desired total depth, and K is how far the tool will drill in on each peck before retracting.

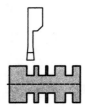

G75 GROOVING CYCLE

Format: N_ G75 X_ Z_ F_ D_ I_ K_

X   Diameter of groove
Z   Z position of groove
F   Incremental retract
D   Depth/X offset
K   Z movement
I   X movement

The G75 command is used to machine grooves (Fig. 6.15).

EXAMPLE: N25 G75 X0.25 Z-0.75 F0.125 I0.125 K0.125

This command defaults to the last specified feedrate. The F address is used to specify the retract distance, so the feedrate cannot be set within the grooving cycle.

Sample Program G75EX75:

Workpiece Size: Length 4", Diameter 1"

Tool: Tool #5, Grooving Tool

Tool Start Position: X2, Z3

```
%
:1075
N5 G90 G20
N10 T0505 F0.015
N15 M03
N20 G00 Z-0.5 M08
N25 X1.2
```
N30 G75 X0.5 Z-0.75 F0.125 D0 I0.125 K0.125 (Grooving
cycle)
```
N35 G00 X4
N40 Z3 M09
N45 T0500 M05
N50 M30
```

FIGURE 6.15
A grooving example.

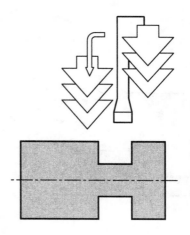

G76 THREADING CYCLE

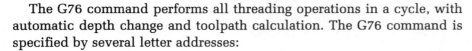

Format: N_ G76 X_ Z_ D_ K_ A_ F_

The G76 command performs all threading operations in a cycle, with automatic depth change and toolpath calculation. The G76 command is specified by several letter addresses:

X	Minor diameter of thread
Z	Position at end of thread
D	Depth of first pass in thousandths (the control uses the first depth of cut to determine the number of passes)
K	Depth of thread
F	Pitch of thread (thread pitch = 1/thread/in.)
A	Tool angle (If the tool angle is given, the tool will continue to cut on the leading edge of the tool; if no tool angle is given, the tool will cut on both sides.)

EXAMPLE: N25 G76 X0.5 Z-1.0 D625 K0.25 A55 F0.1

In this example, the tool cuts a thread, starting at its present location and ending at the specified XZ endpoint. The D value specifies each cut depth, and the K value defines the overall depth. The A value defines the tool angle, and the F value defines the pitch (Fig. 6.16).

Sample Program G76EX76:

Workpiece Size: Length 4", Diameter 1"

Tool: Tool #6, Neutral Tool

Tool Start Position: X2, Z3

```
%
:1076
N5 G90 G20
N10 T0606 M08
N15 M03
N20 G00 X1
N25 Z0.1
```

(continues)

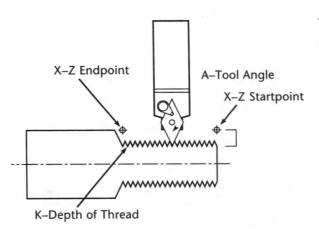

X–Z Endpoint

A–Tool Angle

X–Z Startpoint

K–Depth of Thread

FIGURE 6.16
Startpoint and endpoint of thread cycle.

(continued)

N25 G76 X0.96 Z-2 D625 K0.125 A55 F0.1 (Threading cycle)

N30 G00 X4

N35 Z3 M09

N35 T0600 M05

N40 M30

G90 ABSOLUTE POSITIONING

Format: N_ G90

The G90 command sets the system to accept all coordinates as absolute data and can be switched only with the G91 command.

EXAMPLE: N15 G90

In this example, any coordinate data entered after this command are absolute coordinates.

Sample Program G90EX90:

Workpiece Size: Length 4", Diameter 2"

Tool: Tool #1, Right-hand Tool

Tool Start Position: X2, Z3

```
%
:1090
```

N5 G90   (Input all data in absolute mode)

```
N10 G20
N15 T0101
N20 M03
N25 G00 X1.75 M08
N30 Z0.1
N35 G01 Z-2 F0.015
N40 G00 X2 Z0.1
N45 X1.5
N50 G01 Z-1.75
N55 G00 X1.75 Z0.1
N60 X1.25
N65 G01 Z-1.5
N70 G00 X1.5 Z0.1
N75 X1
N80 G01 Z-1.25
N85 X2 Z-2.25
N90 G00 X4 Z3 M09
N95 T0100 M09
N100 M30
```

G91 INCREMENTAL POSITIONING

Format: N_ G91

The G91 command sets the system to accept all coordinates as incremental data and can be switched with the G90 command within a program.

EXAMPLE: N25 G91

In this example, any coordinates entered after this command are incremental coordinates.

Sample Program G91EX91:

Workpiece Size: Length 4", Diameter 2"

Tool: Tool #1, Right-hand Tool

Tool Start Position: X2, Z3

```
%
:1091
N5 G90 G20
N10 T0101
N15 M03
N25 G00 X1.75
N30 Z0.1
N35 G91 G01 Z-1.85 F0.015     (Incremental data input)
N40 G00 X0.25 Z1.85
N45 X-0.5
N50 G01 Z-1.6
N55 G00 X0.25 Z1.6
N60 X-0.5
N65 G01 Z-1.35
N70 G00 X0.25 Z1.35
N75 X-0.5
N80 G01 Z-1.1
N85 X1 Z-1
N90 G90 G00 X4 Z3 M09
N95 T0100 M05
N100 M30
```

G96 CONSTANT SURFACE SPEED SETTING

Format: N_ G96 S_

When constant surface speed (CSS) is set, the spindle speed will vary according to the tool position in the X direction. To maintain CSS the MCU automatically calculates the appropriate spindle speed to maintain proper relative speed between the tool and the workpiece. As the tool approaches the center of the rotating workpiece, the spindle speed will increase. The surface speed depends on the input units setting. If it is in metric units, the surface speed is set to meters per minute; if it is in inches, the surface speed is set to feet per minute. Note that the CSS is inversely proportional to the radius.

Sample Program G96EX96:

Workpiece Size: Length 4", Diameter 2"

Tools: Tool #10, 1" Drill

    Tool #12, Center Drill

Tool Start Position: X2, Z3

```
%
:1096
N5 G20 G40 G99
N10 T1212 M08
N12 G96 S600                (Set CSS at 600 feet per minute)
N15 G98 M03 S2000
N20 G00 X0
N25 Z0.1
N30 G01 Z-0.2 F2
N35 G00 Z3
N40 X4 T1200
N45 T1010
N50 G00 X0
N55 Z0.1
N60 G74 Z-1 F0.5 D0 K0.25
N65 G99 G00 Z3
N70 T1000 X4
N75 M05
N80 M30
```

G97 CONSTANT SPINDLE SPEED

Format: N_ G97 S_

The G97 command is used to cancel the G96 setting and set the spindle speed in revolutions per minute. The tool position does not affect the spindle speed.

Sample Program G97EX97:

Workpiece Size: Length 4", Diameter 2"

Tools: Tool #10, 1" Drill

    Tool #12, Center Drill

Tool Start Position: X2, Z3

```
%
:1097
N5 G20 G40 G99
N10 T1212 M08
N12 G97 S800         (Set the spindle speed at 800 rev. per minute)
N15 G98 M03 S2000
N20 G00 X0
```

```
N25 Z0.1
N30 G01 Z-0.2 F2
N35 G00 Z3
N40 X4 T1200
N45 T1010
N50 G00 X0
N55 Z0.1
N60 G74 Z-1 F0.5 D0 K0.25
N65 G99 G00 Z3
N70 T1000 X4
N75 M05
N80 M30
```

G98 LINEAR FEEDRATE PER TIME

Format: N_ G98

The G98 command sets the linear feedrate to units (inch or millimeter) per time (minutes). The setting depends on which type of unit is currently active.

It causes all feedmoves (G01, G02, and G03, as well as the canned cycles) to accept units per minute as the defining rate. It sets the feedrate only to units per minute; it does *not* set the actual feedrate for the feedmoves.

Sample Program G98EX98:

Workpiece Size: Length 4", Diameter 2"

Tools: Tool #10, 1" Drill

                            Tool #12, Center Drill

Tool Start Position: X2, Z3

```
%
:1041
N5 G20 G40 G99
N10 T1212 M08
N15 G98 M03 S2000        (Set feed in inches per minute)
N20 G00 X0
N25 Z0.1
N30 G01 Z-0.2 F2
N35 G00 Z3
N40 X4 T1200
N45 T1010
N50 G00 X0
N55 Z0.1
N60 G74 Z-1 F0.5 D0 K0.25
N65 G99 G00 Z3            (Set feed in inches per revolution)
N70 T1000 X4
N75 M05
N80 M30
```

G99 FEEDRATE PER REVOLUTION

Format: N_ G99

The G99 command sets the feedrate to units (inch or millimeter) per revolution. The setting depends on which type of unit is currently active.

It causes all feedmoves (G01, G02, and G03, as well as the canned cycles) to accept units per revolution as the defining rate. It is the default setting on most CNC turning centers. It also is used to synchronize axes moves with the spindle, especially in programming threads.

Sample Program G99EX99:

Workpiece Size:	Length 4", Diameter 2"
Tools:	Tool #8, Left-hand Facing Tool
	Tool #10, 1" Drill
	Tool #12, Center Drill

Tool Start Position: X2, Z3

```
%
:1041
N5 G20 G40 G99            (Set to inches per revolution)
N10 T1212 M08
N15 G98 M03 S2000         (Set to inches per minute)
N20 G00 X0
N25 Z0.1
N30 G01 Z-0.2 F2.0
N35 G00 Z3
N40 X4 T1200
N45 T1010
N50 G00 X0
N55 Z0.1
N60 G74 Z-1 F0.5 D0 K0.25
N65 G99 G00 Z3            (Reset to inches per revolution)
N70 T1000 X4
N75 T0101
N80 G00 X0.75
N85 Z0.1
N90 G01 Z0 F0.015
N95 X1.5
N100 G03 X2 Z-0.25 I0 K-0.25
N105 G01 X1.75 Z-1.5
N110 X2 Z-3
N115 G00 X4 Z3 M09
N120 T0100 M05
N125 M30
```

M-CODES

M-codes are miscellaneous functions that include actions necessary for machining but not those that are actual tool movements (for example, auxiliary functions). They include spindle on and off, tool changes, coolant on and off, program stops, and similar, related functions. The following codes are described in detail in the following sections.

M00	Program stop
M01	Optional program stop
M02	Program end
M03	Spindle on clockwise
M04	Spindle on counterclockwise
M05	Spindle stop
M07	Coolant 1 on
M08	Coolant 2 on
M09	Coolant off
M98	Subprogram call
M99	Return from subprogram
M30	End of program, reset to start
Block Skip	"/" used to bypass CNC blocks
Comments	"(" and ")" used to help operator comments

Note: Two or more M-codes cannot be included on the same line.

M00 PROGRAM STOP

Format: N_ M00

The M00 command is a temporary program stop function. All functions are temporarily suspended and remain so until reactivated by user input.

When the MCU encounters an M00 command, it shuts down all machine controls (spindle, coolant, all axes, and any auxiliaries). The MCU then waits for user input to continue the program. The screen prompt "Program Stop, Enter to Continue" is displayed on screen. The MCU will reinstate all functions only after Enter is pressed.

The MCU retains all program blocks and coordinate data until after the machine control has been returned to its normal state. (The computer remembers all points and tool position information so that there is no chance of losing the program zero.)

EXAMPLE: N250 M00

In this example, the program cycles through until the M00 command is encountered. The machine and all operations then halt at N250 and await user input to continue the program.

Sample Program M00EX0:

Workpiece Size: Length 4", Diameter 2"

Tool: Tool #1, Right-hand Tool

Tool Start Position: X2, Z3

```
%
:1000
N5  T0101
N10 M03
N15 G00 X2.1 M08
N20 Z-0.1
N25 G01 X0 F0.015
N30 G00 X4 Z0
N35 M00                    (Program stop—MCU waits)
N40 X2.1 Z-0.3
N45 G01 X1.4 Z-0.2
N50 X1
N55 Z0.1
N60 G00 Z3
N65 X4 M09
N70 T0100 M05
N75 M30
```

M01 OPTIONAL PROGRAM STOP

Format: N_ M01

The M01 command is an optional program stop that stops the program only if the M01 switch is selected or set to ON.

In the Options menu there is an Optional Stop item. Its default mode is OFF, or unchecked ☐. To turn it on, simply click on it to make the check mark ☑ appear next to the option text. The switch will then be set to ON.

With the M01 switch on, the MCU halts all machine functions temporarily. The monitor displays the screen prompt "Optional Stop, Enter to Continue." After you press Enter, the lathe will carry on as before, without any loss of position.

EXAMPLE: N25 M01

In this example, the machine and all operations halt temporarily at N25, provided the switch is on.

The following sample program is identical to that for the M00 command except that M01 replaces M00. If the M01 switch is not turned on, the program will ignore the command. Try running the program first with the Optional Stop switch off and then with it on.

Sample Program M01EX1:

Workpiece Size: Length 4", Diameter 2"

Tool: Tool #1, Right-hand Tool

Tool Start Position: X2,Z3

Optional Stop: On

```
%
:1000
N5 T0101
N10 M03
N15 G00 X2.1 M08
N20 Z-0.1
N25 G01 X0 F0.015
N30 G00 X4 Z0
N35 M01                    (Optional program stop)
N40 X2.1 Z-0.3
N45 G01 X1.4 Z-0.2
N50 X1
N55 Z0.1
N60 G00 Z3
N65 X4 M09
N70 T0100 M05
N75 M30
```

M02 PROGRAM END

END

Format: N_ M02

The M02 command indicates the end of the main program cycle operation. Upon encountering this command, the MCU switches off all machine operations (spindle, coolant, all axes, and any auxiliaries) and terminates the program.

The M02 command is always by itself on the last line of the program.

Sample Program M02EX2:

Workpiece Size: Length 4", Diameter 2"

Tools: Tool #1, Right-hand Tool

                    Tool #2, Right-hand Finishing Tool

Tool Start Position: X2, Z3

```
%
:1070
N5 G90 G20
N10 T0101
N15 M03
N20 G00 X2 Z0.1 M08
N35 G71 P40 Q60 U0.05 W0.05 D625 F0.012
N40 G01 X0 Z0
N45 G03 X0.5 Z-0.25 I0 K-0.25
N50 G01 X1.5 Z-0.75
N55 Z-1.25
N60 X2.1
```

(continues)

(continued)

```
N65  T0100 G00 X4 Z3
N70  T0202
N75  G00 X2 Z0.1
N80  G70 P40 Q60 F0.006
N85  G00 X2.125 Z-1.5
N90  G01 X2 F0.012
N95  G02 Z-3 R2
N100 G00 X4 Z3 M09
N90  T0200 M05
N95  M02                    (Program end)
```

M03 SPINDLE ON CLOCKWISE

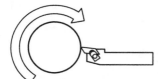

Format: N_ M03 S_

The M03 command switches the spindle on so that it rotates clockwise. The spindle will rotate clockwise at the specified speed (rpm) until instructed by the MCU to stop. You can set the rpm by entering an S code followed by the desired value (for example, S2000). Otherwise, the MCU will determine the rpm based on the desired feedrate per revolution.

EXAMPLE: M03 S2000

In this example, the spindle is turned on at an initial rate of 2000 rpm.

Sample Program M03EX3:

Workpiece Size: Length 100 mm, Diameter 50 mm

Tool: Tool #3, Neutral Tool

Tool Start Position: X50, Z75

```
%
:1021
N5  G90 G99 G21 G40
N15 T0303
N17 M03                 (Spindle on clockwise)
N20 G00 Z2 M08
N25 X45
N30 G01 Z-60 F0.015
N35 X40
N40 Z-10
N45 X35
N50 Z-25
N55 X30
N60 Z-10
N65 G00 X100 M09
N70 Z100
N75 T0300 M05           (Spindle off)
N80 M30
```

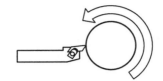

M04 SPINDLE ON COUNTERCLOCKWISE

Format: N_ M04 S_

The M04 command switches the spindle on so that it rotates counter-clockwise. The M04 command is specified by the letter address S and the required rpm.

The spindle will rotate counterclockwise at the specified speed (rpm) until instructed by the MCU to stop. You can change the spindle speed at any point during the program by redefining the S value.

EXAMPLE: M04 S2000

In this example, the spindle is turned on at an initial rate of 2000 rpm.

Sample Program M04EX4:

Workpiece Size: Length 4", Diameter 2"

Tool: Tool #2, Right-hand Finishing Tool

Tool Start Position: X2, Z3

```
%
:1002
N5 G20 G99 G40
N10 T0202
N15 M04                     (Spindle on counterclockwise)
N20 G00 X1.7
N22 Z0.1 M08
N25 G01 Z-0.5 F0.012
N30 G00 X2
N35 Z0.1
N40 X1.4
N45 G01 Z-0.25
N50 G00 X1.5 Z0.1
N55 X1
N60 G01 Z-0.1
N65 G00 X1.1 Z0.1
N70 X0
N75 G01 Z0
N80 G03 X2 Z-1 I0 K-1
N85 G00 X4 Z3 M09
N90 T0200 M05               (Spindle off)
N95 M30
```

M05 SPINDLE STOP

Format: N_ M05

The M05 command stops the spindle rotation. Although the M00, M01, and M02 commands switch the spindle off temporarily, only the M05 command switches the spindle off directly.

EXAMPLE: N350 M05

In this example, the machine spindle stops at N350.

Sample Program M05EX5:

Workpiece Size: Length 4", Diameter 2.5"

Tool: Tool #1, Right-hand Tool

Tool Start Position: X2, Z3

```
%
:1001
N5 G20 G40
N10 T0101
N15 M03
N20 G00 X2.375 M08
N22 Z0.1
N25 G01 Z-2.0 F0.015
N30 G00 X2.5
N35 Z0.1
N40 X2.25
N45 G01 Z-2
N50 G00 X2.375
N55 Z0.1
N60 X2.125
N65 G01 Z-2
N70 G00 X2.25
N75 Z0.1
N80 X1.875
N85 G01 Z0
N90 X2.125 Z-0.125
N95 G00 X4 M09
N100 Z3
N105 T0100 M05          (Turn spindle off)
N110 M30
```

M07 COOLANT 1 ON

Format: N_ M07

The M07 command switches on the first coolant flow. Coolant is required when you are turning materials such as mild steel. It provides lubrication, cools the tool, and carries away some of the chips.

When you are machining lighter materials such as aluminum or wax, coolant is not always required.

EXAMPLE: N475 M07

In this example, the coolant flow from the number one hose is turned on when the MCU encounters the M07 command.

The status of this command can be verified by inspecting the status window for the coolant function. M07 is turned off by the M09 command after turning has been completed.

Sample Program M07EX7:

Workpiece Size: Length 4", Diameter 2"

Tool: Tool #3, Neutral Tool

Tool Start Position: X2, Z3

```
%
:1021
N5  G90  G20  G40
N15  T0303
N17  M03
N20  G00  X2.1  M07          (Coolant Hose #1 on)
N25  Z-1.25
N30  G01  X1.75  F0.015
N35  Z-2.75
N40  X1.5
N45  Z-1.25
N50  G02  Z0  R1  F0.006
N55  G00  X4  Z3  M09        (Coolant off)
N60  T0300  M05
N65  M30
```

M08 COOLANT 2 ON

Format: N_ M08

The M08 command switches on the second coolant flow. Coolant is required when you are turning materials such as mild steel. It provides lubrication, cools the tool, and carries away some of the chips.

When you are machining lighter materials such as aluminum or wax, coolant is not always required.

EXAMPLE: N75 M08

In this example, the coolant flow from the number two hose is turned on when the MCU encounters the M08 command.

Sample Program M08EX8:

Workpiece Size: Length 4", Diameter 2"

Tool: Tool #3, Neutral Tool

Tool Start Position: X2, Z3

```
%
:1008
N5  G90  G20  G40
N15  T0303
N17  M03
N20  G00  X2.1  M08          (Coolant Hose #2 on)
N25  Z-1.25
N30  G01  X1.75  F0.015
N35  Z-2.75
N40  Z-2.5
```

(continues)

(continued)

```
N45  X1.5
N50  Z-1.5
N55  G00  X2.1
N60  Z-1
N65  G01  X1.75
N70  Z-0.25
N75  G00  X4
N80  Z3  M09                    (Coolant off)
N85  T0300  M05
N90  M30
```

M09 COOLANT OFF

Format: N_ M09

The M09 command switches off all coolant flow. The coolant is switched off automatically when a tool change, a program end, or a program stop is encountered. You are responsible for switching the coolant off when required. Refer to the status window to verify whether coolant is on or off.

EXAMPLE: N10 M09

In this example, all coolants (from both hose number one and hose number two) are turned off when the MCU encounters the M09 command.

Sample Program M09EX9:

Workpiece Size: Length 4", Diameter 2"

Tool: Tool #3, Neutral Tool

Start Position: X2, Z3

```
%
:1009
N5  G90  G20  G40
N15  T0303
N17  M03
N20  G00  X2.1  M08            (Coolant on)
N25  Z-2.75
N30  G01  X1.75  F0.015
N35  Z-0.25
N40  X1.5
N45  Z-1.5
N50  X1.25
N55  X1.5  Z-0.25
N60  G00  X4
N65  Z3  M09                   (All coolant hoses off)
N70  T0300  M05
N75  M30
```

M30 END OF PROGRAM, RESET TO START

Format: N_ M30

The M30 command stops program execution and awaits user input before continuing the program. It then rewinds the program memory to the beginning. On older CNC or NC machines that use paper tape, this command will rewind the tape to the beginning of the program.

The prompt "Program End, Enter to Continue" is displayed on the screen (as a safety measure to prevent accidental program end).

EXAMPLE: N50 M30

In this example, program execution is stopped at N50.

Sample Program M30EX30:

Workpiece Size: Length 4", Diameter 2"

Tool: Tool #1, Right-hand Tool

Tool Start Position: X2,Z3

```
%
:1002
N5 G20 G40
N10 T0101
N15 M03
N20 G00 Z0.05 X1.75 M08
N25 G01 Z-2.5 F0.015
N30 X2 Z-2.75
N35 G00 X2.1 Z0.05
N40 X1.25
N45 G01 Z0
N50 G03 X1.75 Z-0.25 I0 K-0.25
N55 G00 X4 Z3 M09
N60 T0100 M05
N65 M30                    (End of Program)
```

BLOCK SKIP

Format: / N_

The use of Block Skip is extremely helpful in family of parts programming. This functionality is directly controlled by the Block Skip switch on most CNC Controllers. With the CNCez simulators this option is found in the Options menu. A check mark next to this option item indicates that Block Skip is selected, or turned on. If turned on, upon execution of a CNC program and a "/" is encountered, the program will ignore any CNC code on that block.

An example of using Block Skip would be when two customers want a similar part machined with minor differences. The part for customer A may require several different machining operations or hole patterns, but the part for customer B may not require a particular pattern. The machine operator would simply edit the CNC program and insert Block Skips in front of the CNC blocks that are not required for customer B. He can then run both jobs anytime simply by selecting the Block Skip switch on the MCU front panel.

When running the following sample, first try it with Block Skip on, then repeat the program with Block Skip off.

Sample Program SKIPEX2:

Workpiece Size: X2.5, Z3

Tool: Tool #1, Center Drill

                    Tool #2, 3/8" Drill

                    Tool #3, 1/2" Drill

                    Tool #4, Inside Turning Tool

                    Tool #5, R. H. Turning Tool

                    Tool #6, Grooving/Parting Tool

Tool Start Position: X2, Z3

```
%
:1001
N5 G90 G98 G20 G40
N10 G00 X0 Z.25
N15 T0101
N20 M03 S850
N25 G74 Z-.25 K0.125 F2
N30 T0100
N35 T0202
N40 G00 X0 Z.25
N45 G74 Z-1 K0.125 F2
N50 T0200
N55 T0303
N60 G00 X0 Z.25
N65 G74 Z-1.5 K0.125 F2
N70 T0300
/N75 T0404
/N80 G00 X.675 Z.2
/N85 G01 Z-1.125 F2
/N90 X.75
/N95 Z-1
/N100 X1.25 Z-.75
/N105 Z-.5
/N110 X1.25 Z-.3
/N115 Z.1
/N120 G00 Z.5
/N125 T0400
N130 T0505
N135 G00 X2.55
N140 Z.2
N145 X2.25
```

```
N150 G01 Z-2.25 F3
N155 X2.5 Z-2
N160 G00 Z.2
N165 T0500
N170 T0606
N175 G00 X2.75
N180 Z-2.0
N185 G75 X1.5 Z-2.25 F0.125 I0.125 K0.125
N187 G00 Z2
N190 T0600
N195 M05
N200 M30
```

COMMENTS

Format: N_   (Comment statement)

Comments are useful to the CNC machine operator when setting up and running a job. Comments are defined by the use of round brackets. Anything between them is ignored by the program. Throughout the examples, comments are used to help you understand the CNC codes. Remember that comments are just aids in reading and understanding a program; the text is totally ignored even if it contains valid CNC code.

STEP-BY-STEP TURNING EXAMPLES

Work through each of the following examples step by step. In the first example, each step is described in more detail than it is in the remaining examples, so be sure to work through it.

If you have any difficulties with the programs or simply want to test and see them without entering them into the editor, they can be found in the Demoturn folder.

EXAMPLE 1:   I-turn1.trn

This program introduces you to the basic G- and M-codes. When executed, it turns the workpiece from an original diameter of 1.0 in. The completed part is shown in Fig. 6.17.

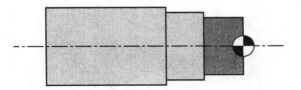

FIGURE 6.17
The completed part.

Workpiece Size: 1" Diameter by 2" Length

Tool: Tool #1, Right-hand Turning Tool

Tool Start Position: X2, Z3

```
%
:1001
```

(continues)

(continued)

```
N5   G20 G40
N10  T0101 M03
N15  G00 X1 Z0.25 M08
N20  G01 Z-0.75 F0.015
N25  G00 X1.1
N30  Z0.05
N35  X0.9375
N40  G01 Z-0.75 F0.012
N45  G00 X1
N50  Z0.05
N55  X0.8745
N60  G01 Z-0.75
N65  G00 X0.5
N70  Z0.05
N75  X0.8125
N80  G01 Z-0.375
N85  G00 X1
N90  Z0.05
N95  X0.75
N100 G01 Z-0.375
N105 G00 X2 M09
N110 Z2
N115 T0100 M05
N120 M30
```

STEP 1: Run the CNCez Turning Simulator either from the Windows taskbar or Desktop icon, or from the CNC Workshop CBT. (Fig. 6.18)

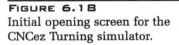

FIGURE 6.18
Initial opening screen for the CNCez Turning simulator.

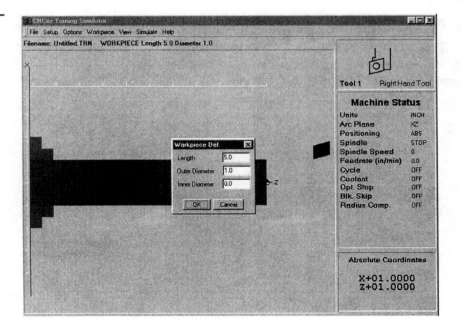

STEP 2: The CNCez graphic user interface for Turning is displayed (Fig. 6.19).

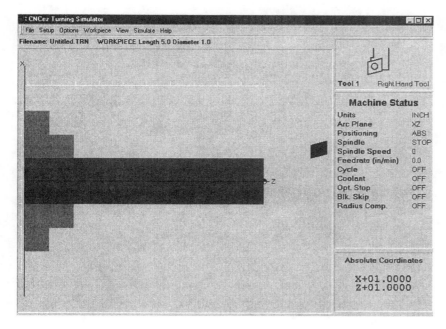

FIGURE 6.19
The graphic user interface for turning.

STEP 3: Create a new file called I-turn1.

Move the pointer to the menu bar and select File.

Select New... (Fig. 6.20).

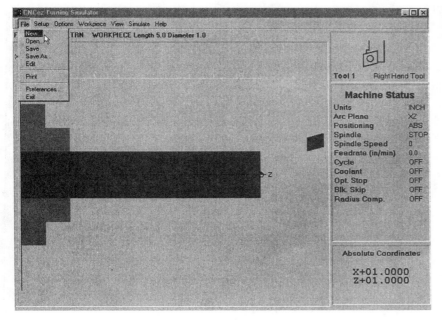

FIGURE 6.20
Selecting New... from the File menu.

Enter the file name: I-turn1.

Click on OK or [Enter] (Fig. 6.21).

FIGURE 6.21
Dialog to enter the filename.

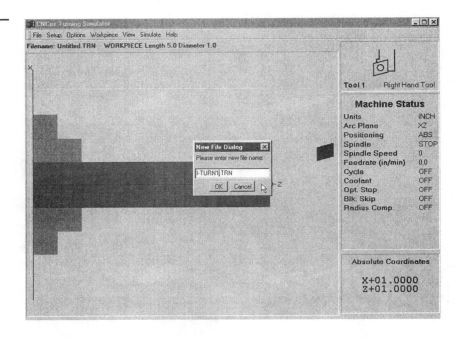

FIGURE 6.21
Dialog to enter the filename.

STEP 4: Set up the workpiece (stock material) for this program.

From the menu bar, select Workpiece.

Select New... (Fig. 6.22).

FIGURE 6.22
Selecting New...from the
Workpiece menu.

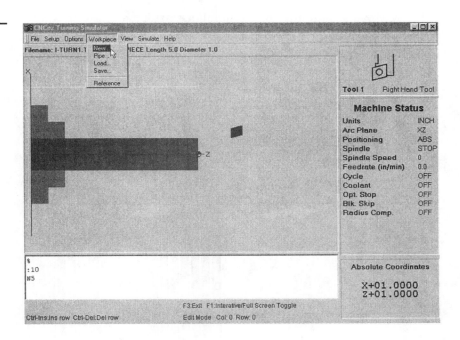

Enter the length (in.): 2.

Enter the diameter (in.): 1.

Click on OK (Fig. 6.23).

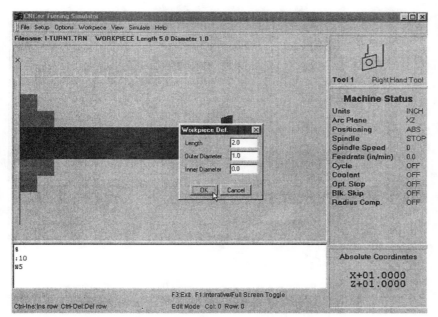

FIGURE 6.23
Entering the workpiece dimensions in the Workpiece Definition dialog.

STEP 5: This CNC program will use the default Tool Library that is shipped with the software. The program uses Tool #1 which is defaulted in the turret to be a right-hand tool. Look in the Tool Library at Tool #1.

From the menu bar, select Setup.

Select Tools (Fig. 6.24). The Tool Setup dialog will appear.

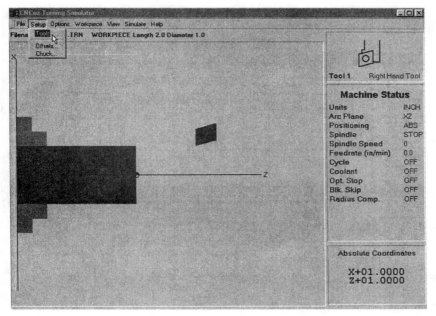

FIGURE 6.24
Selecting Tools... from the Setup menu.

Move the pointer over Tool #1 of the Tool Library on the left; clicking on it will change the green box to red. This indicates that the tool has been selected (Fig. 6.25).

FIGURE 6.25
Selecting Tool #1 from the Tool
Library.

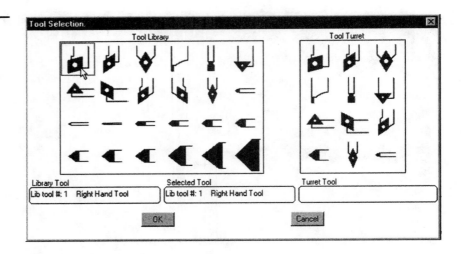

Click on Tool Turret Location #1 to make this the right-hand
tool (Fig. 6.26).

Click on OK to accept these changes to the Tool Turret.

FIGURE 6.26
Selecting the Tool Turret location
#1.

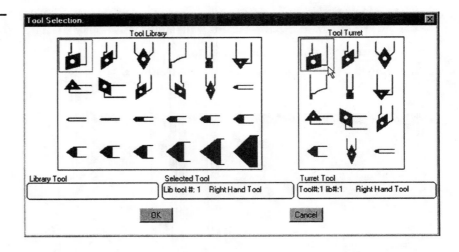

STEP 6: Begin entering the program and simulate the cutter path. If
you are not in the Simulate/Edit mode, use the Simulate/Edit
option.

From the menu bar, select Simulate.

Select Edit (Fig. 6.27)

STEP 7: Program setup phase. You must enter all the setup
parameters before you can enter actual cutting moves (Figs.
6.28, 6.29, and 6.30).

>% [Enter]   Program start flag

>:1001 [Enter]   Program number 1001 (Fig. 6.28)

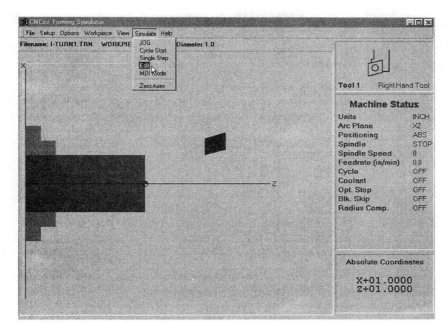

FIGURE 6.27
Selecting Edit from the Simulate menu.

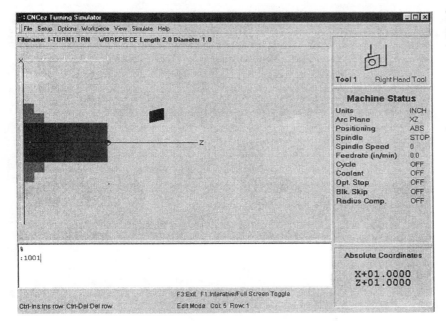

FIGURE 6.28
Entering the program number 1001.

>:N5 G20 G40 [Enter]	Inch input and TNR cancel
>:N10 T0101 M03 [Enter]	Tool #1, spindle on clockwise (Fig. 6.29)

STEP 8: Material removal phase. Begin cutting the workpiece, using G00 and G01.

>N15 G00 X1 Z0.25 M08 [Enter]	Rapiding to (X1, Z0.25) (Fig. 6.30)

FIGURE 6.29
Result from entering T0101 M03.

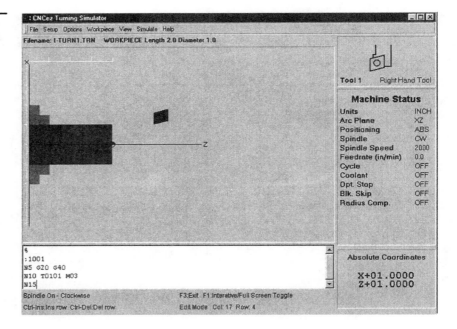

FIGURE 6.30
Rapiding to X0.5 Z0.25.

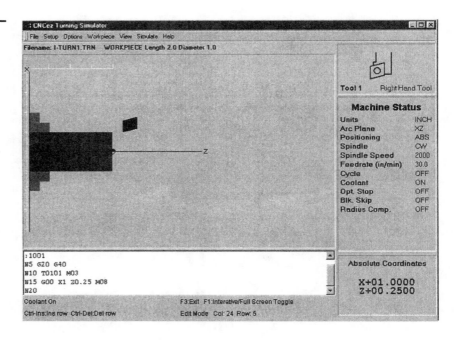

>N20 G01 Z-0.75 F0.015 [Enter]   G01 feed move to Z-0.75 at 0.015 IPR (Fig. 6.31)

>N25 G00 X1.1 [Enter]   G00 up to X1.1

>N30 Z0.05 [Enter]   G00 to Z0.05

>N35 X0.9375 [Enter]   G00 down to X0.9375

>N40 G01 Z-0.75 F0.012 [Enter]   G01 feed move to Z-0.75 at 0.012 ipr (Fig. 6.32)

>N45 G00 X1 [Enter]   G00 up to X1

>N50 Z0.05 [Enter]   G00 right to Z0.05

>N55 X0.8745 [Enter]   G00 down to X0.8745

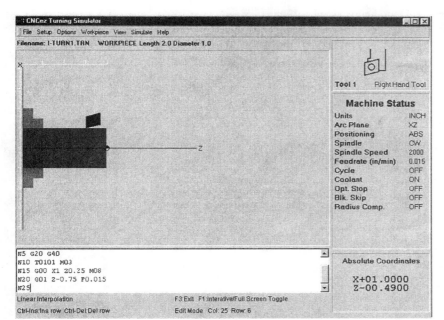

FIGURE 6.31
Performing a G01 to Z-0.75.

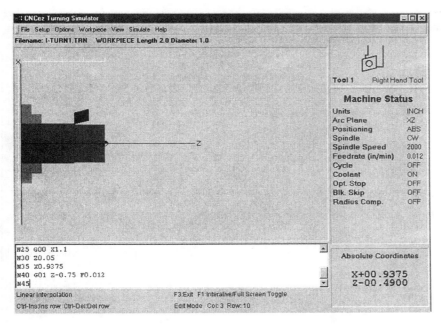

FIGURE 6.32
Performing another G01 to
Z-0.75 on N40.

>N60 G01 Z-0.75 [Enter]	G01 feed move to Z-0.75 at 0.012 ipr (Fig. 6.33)
>N65 G00 X1 [Enter]	G00 up to X1
>N70 Z0.05 [Enter]	G00 right to Z0.05
>N75 X0.8125 [Enter]	G00 down to X0.8125
>N80 G01 Z-0.375 [Enter]	G01 feed move to Z-0.375 at 0.012 ipr (Fig. 6.34)
>N85 G00 X1 [Enter]	G00 rapid to X1
>N90 Z0.05 [Enter]	G00 rapid to Z0.05
>N95 X0.75 [Enter]	G00 rapid to X0.75

FIGURE 6.33
Performing another G01 to
Z-0.75 on N60.

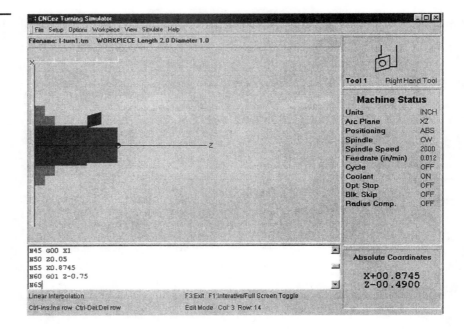

FIGURE 6.34
Performing another G01 to
Z-0.75 on N80.

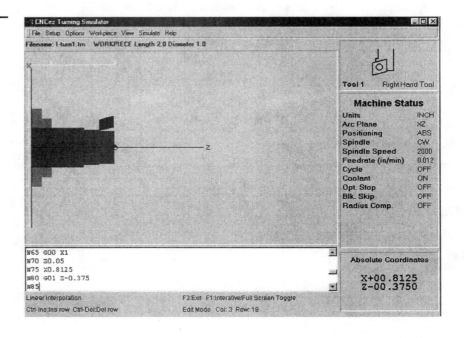

>N100 G01 Z-0.375 [Enter]   G01 feed move to Z-0.375

>N105 G00 X2 M09 [Enter]   G00 rapid to X1 and coolant off

>N110 Z2 [Enter]   G00 rapid to Z2

STEP 9: Program shutdown phase. Turn off the spindle and program
end (Fig. 6.35).

>N115 T0100 M05 [Enter]   M05 spindle off

>N120 M30 [Enter]   End of program (Fig. 6.35)

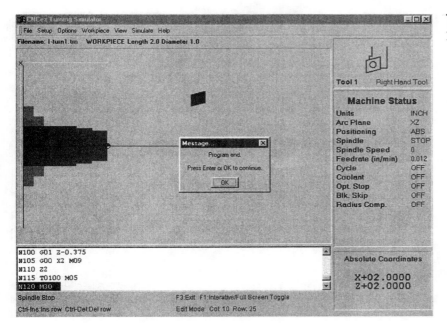

FIGURE 6.35
End of program dialog prompt.

>N65 Press F3   Exit from Simulate/Edit mode

STEP 10: The program is complete. Rerun it, using the Simulate/Cycle option.

From the menu bar, select Simulate.

Select Cycle Start (Fig. 6.36).

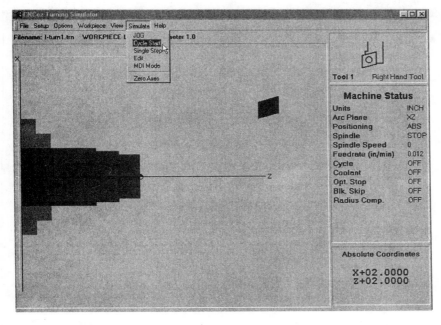

FIGURE 6.36
Selecting Cycle Start from the Simulate menu.

FIGURE 6.37
End of program.

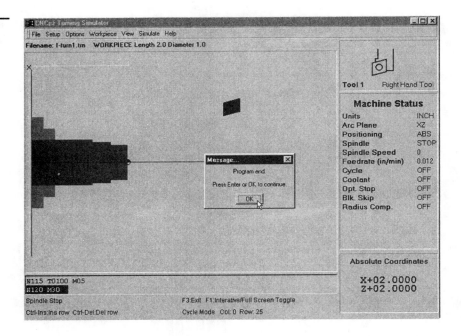

STEP 11: End of program (Fig. 6.37)

STEP 12: View the completed solid.

From the menu bar, select View.

Select Solid (Fig. 6.38)

FIGURE 6.38
Selecting the View Solid option.

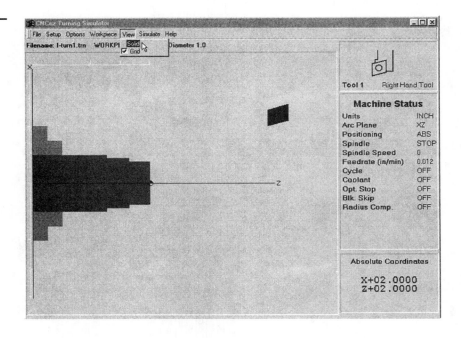

A Solid View dialog will be displayed. Here you can rotate the workpiece and obtain a hard copy printout as well (Fig. 6.39).

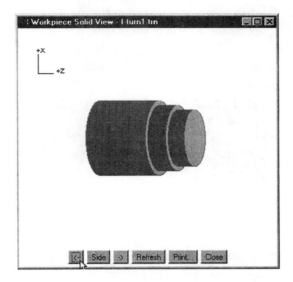

FIGURE 6.39
The solid view representation of
the completed part.

STEP 13: You can optionally save this program to a working folder, say, "WORKTURN."

EXAMPLE 2: I-turn2.trn

This program introduces you to the basic circular interpolation routines. The completed part is shown in Fig. 6.40.

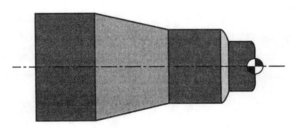

FIGURE 6.40
The completed part.

Workpiece Size: 2" Diameter by 4" Length

Tool: Tool #1, Right-hand Turning Tool

Tool Start Position: X2, Z3

```
%
:1002
N5 G20 G40
N10 T0101
N15 M03
N20 G00 X1.8 Z0.05
N25 M08
N30 G01 Z-2.5 F0.015
N35 G00 X2
N40 Z0.05
N45 X1.6
```

(continues)

(continued)

```
N50   G01  Z-2
N55   G00  X1.8
N60   Z0.05
N65   X1.4
N70   G01  Z-1.5
N75   X2  Z-3
N80   G00  Z0.05
N85   X1.2
N90   G01  Z-0.5
N95   G00  X1.4
N100  Z0.05
N105  X1
N110  G01  Z-0.5
N115  G03  X1.4  Z-0.7  I0  K-0.2
N120  G00  Z-0.1
N125  X1
N135  G02  X0.8  Z0  I-0.1  K0
N140  G00  X2  Z2
N145  M09
N150  T0100  M05
N155  M30
```

STEP 1: Create a new file called I-turn2.

Move the pointer to the menu bar and select File.

Select New...

Enter the file name: I-turn2

Click on OK or [Enter].

STEP 2: Set up the workpiece (stock material) for this program.

From the menu bar, select Workpiece.

Select New...

Enter the length (in.): 4.

Enter the diameter (in.): 2.

Click on OK.

STEP 3: Begin entering the program and simulate the cutter path. Use the Simulate/Edit option. If not in the Simulate/Edit mode, from the menu bar, select Simulate.

Select Edit.

STEP 4: Program setup phase. You must enter all the setup parameters before you can enter actual cutting moves.

>% [Enter]	Program start flag
>:1002 [Enter]	Program number 1002
>:N05 G20 G40 [Enter]	Inch programming and TNR cancel
>:N10 T0101 [Enter]	Tool change to Tool #1
>N15 M03 [Enter]	(Fig. 6.41)

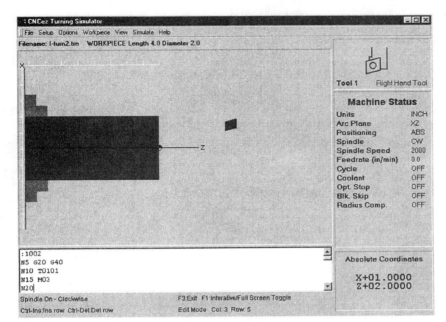

FIGURE 6.41
Spindle on.

STEP 5:   Material removal phase. Begin cutting the workpiece.

>N20 G00 X1.8 Z0.05 [Enter]   Rapiding to (X1.8, Z0.05)

>N25 M08 [Enter]   Coolant on (Fig. 6.42)

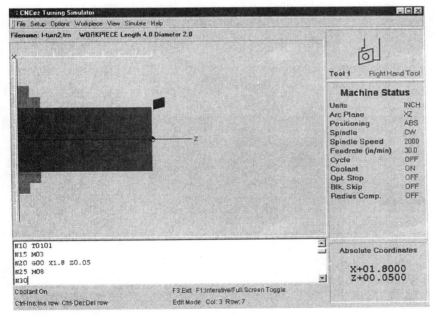

FIGURE 6.42
Tool in position to begin cutting
with coolant on.

>N30 G01 Z-2.5 F0.015 [Enter]   G01 feed move to Z-2.5 at 0.015
                                     ipr (Fig. 6.43)

>N35 G00 X2 [Enter]   G00 rapid to X2

>N40 Z0.05 [Enter]   G00 rapid to Z0.05

FIGURE 6.43
The first cut.

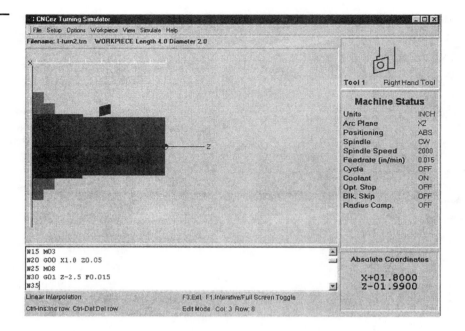

>N45 X1.6 [Enter]	G00 rapid to X1.6
>N50 G01 Z-2 [Enter]	G01 feed move to Z-1
>N55 G00 X1.8 [Enter]	G00 rapid to X1.8
>N60 Z0.05 [Enter]	G00 rapid to Z0.05
>N65 X1.4 [Enter]	G00 rapid to X1.4
>N70 G01 Z-1.5 [Enter]	G01 feed move to Z-1.5
>N75 X2 Z-3 [Enter]	G01 diagonal feed move to (X2, Z-3) (Fig. 6.44)
>N80 G00 Z0.05 [Enter]	G00 rapid to Z0.05

FIGURE 6.44
Tool turning the tapered portion.

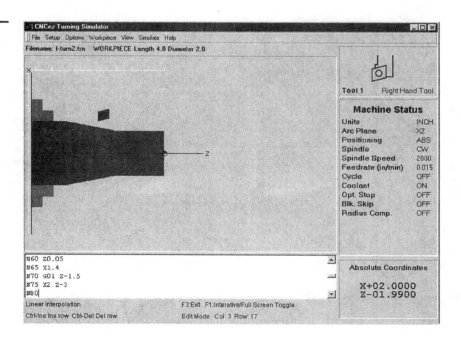

>N85 X1.2 [Enter]   G00 rapid to X1.2

>N90 G01 Z-0.5 [Enter]   G01 feed move to Z-0.5

>N95 G00 X1.4 [Enter]   G00 rapid to X1.4

>N100 Z0.05 [Enter]   G00 rapid to Z0.05

>N105 X1 [Enter]   G00 rapid to X1

>N110 G01 Z-0.5 [Enter]   G01 feed move to Z-0.5 (Fig. 6.45)

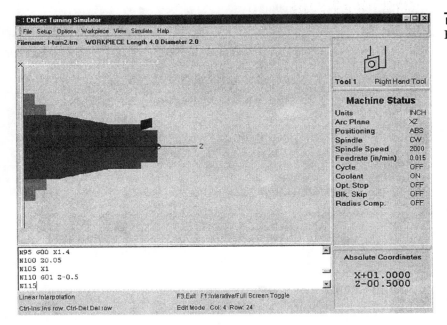

FIGURE 6.45
Result of G01 feed move.

>N115 G03 X1.4 Z-0.7 I0 K-0.2 [Enter]   G03 circular interpolation counterclockwise (Fig. 6.46)

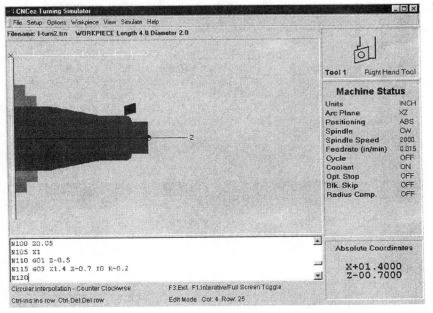

FIGURE 6.46
G03 arc counterclockwise.

>N120 G00 Z-0.1 [Enter]   Rapid move to Z0.1

>N125 X1.1 [Enter]   Rapid move to X1

>N130 G01 X1 [Enter]   Feed move to X1

>N135 G02 X0.8 Z0 I-0.1 K0 [Enter]   G02 arc clockwise (Fig. 6.47)

FIGURE 6.47
Result of G02 arc move.

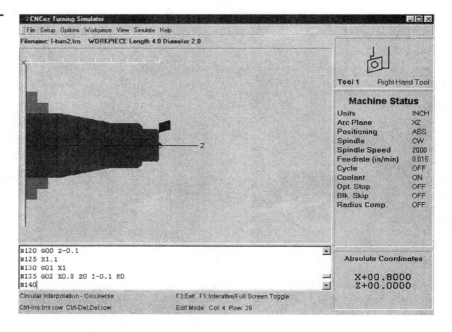

>N140 G00 X2 Z2 [Enter]   Rapid to (X2, Z2)

>N145 M09 [Enter]   Coolant off

>N150 T0100 M05 [Enter]   Spindle off

>N155 M30 [Enter]   End of program (Fig. 6.48)

FIGURE 6.48
End of program dialog prompt.

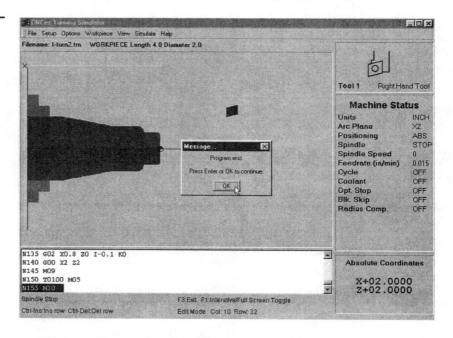

STEP 6:   You can optionally save your part program to your working folder.

EXAMPLE 3:   I-turn3.trn

This program introduces you to the G71 turning cycle and G70 finishing cycle commands. The completed part is shown in Fig. 6.49.

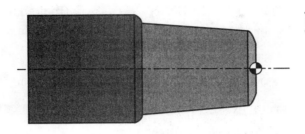

FIGURE 6.49
The completed part.

Workpiece Size: 4" Diameter by 2" Length

Tool: Tool #1, Right-hand Turning Tool

Tool Start Position: X2, Z3

```
%
:1003
N5 G90 G20 G40
N10 T0101
N15 M03
N20 G00 X2.1 Z0.05
N25 G71 P30 Q50 U0.025 W0.005 D625 F0.012
N30 G01 X1 Z0
N35 G03 X1.5 Z-0.25 I0 K-0.25
N40 G01 X1.75 Z-2
N45 G03 X2 Z-2.125 I0 K-0.125
N50 G01 X2.2
N55 G70 P30 Q50 F0.006
N60 G00 X2 Z2
N65 T0100 M05
N70 M30
```

STEP 1:   Create a new file called I-turn3.

Move the pointer to the menu bar and select File.

Select New...

Enter the file name: I-turn3 [Enter] or Click on OK.

STEP 2:   Set up the workpiece (stock material) for this program.

From the menu bar, select Workpiece.

Select New...

Enter the length (in.): 4.

Enter the diameter (in.): 2.

STEP 3:   Begin entering the program and simulate the cutter path. Use the Simulate/Edit option.

STEP 4: Program setup phase. You must enter all the setup parameters before you can enter actual cutting moves.

>% [Enter]	Program start flag
>:1003 [Enter]	Program number 1003
>N5 G90 G20 G40 [Enter]	Absolute, inch, and TNR off
>N10 T0101 [Enter]	Tool change to Tool #1
>N15 M03 [Enter]	Spindle on clockwise

STEP 5: Material removal phase.

>N20 G00 X2.1 Z0.05 [Enter]   Fig. 6.50

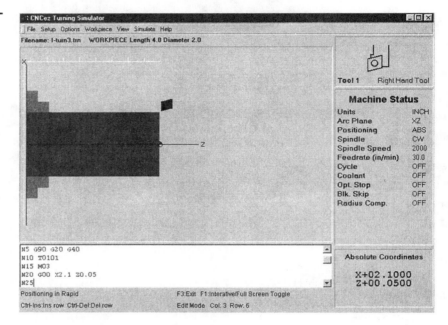

FIGURE 6.50
Tool at cycle ready position.

Before entering the G71 and G70 commands, you must first create the profile. The profile will begin on N30 and end on N50. (This is just a simulation, so don't worry about the depth of cut right now. It will be taken care of by the G71 command.)

>N25 G00 X1 Z0 [Enter]	A temporary line that will later be replaced by the G71 command
>N30 G01 X1 Z0 [Enter]	Feed move to (X1, Z0)
>N35 G03 X1.5 Z-0.25 I0 K-0.25 [Enter]	Arc feed move
>N40 G01 X1.75 Z-2 [Enter]	Fig. 6.51
>N45 G03 X2 Z-2.125 I0 K-0.125 [Enter]	Arc feed move
>N50 G01 X2.2 [Enter]	Fig. 6.52
>N55 G00 X1 Z0 [Enter]	A temporary line that will be replaced by the G70 command
>N60 G00 X2 Z2 [Enter]	Rapid move to (X2, Z2)

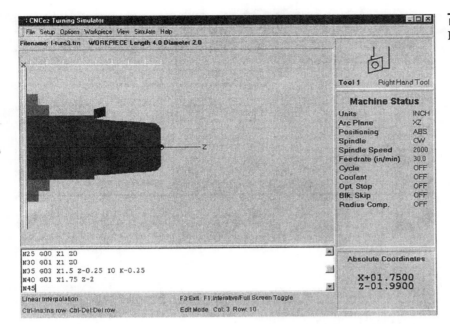

FIGURE 6.51
Result after N40.

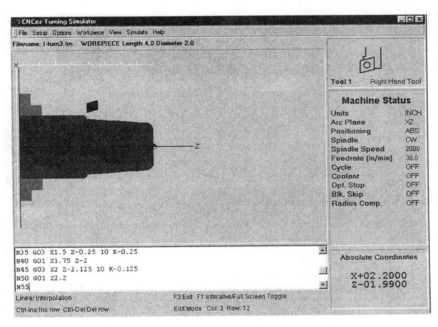

FIGURE 6.52
Result after N50.

>N65 T0100 M05 [Enter]   Spindle off

>N70 M30 [Enter]   Fig. 6.53

STEP 6: Use the File/Edit option to insert the G71 and G70 commands. (Because of how the Simulate/Edit option works, it is much easier and faster to go to the File/Edit option to insert the commands.)

To exit the current mode, press F3.

From the menu bar, select File.

From the File menu, select Edit (Fig. 6.54 and Fig. 6.56).

FIGURE 6.53
The completed profile.

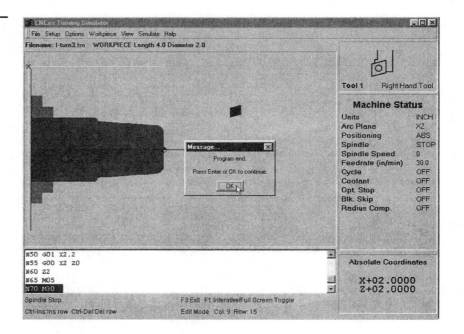

FIGURE 6.54
Selecting Edit from the File menu.

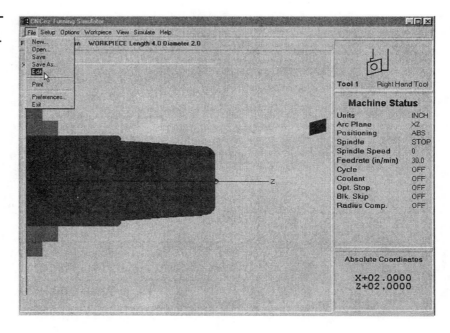

As shown in Figs. 6.55 and 6.56, the editor can be displayed in two ways. The partial view is the way the editor first looks on the screen. When you press F1, the full screen appears.

To select the full-screen editor, press F1, (Fig. 6.56).

Here you must replace N25 and N55 with the following, also shown in Fig. 6.56:

>N25 G71 P30 Q50 U0.025 W0.005 D625 F0.012   Turning cycle

>N55 G70 P30 Q50 F0.006   Finishing cycle

>To exit from the editor mode, Press F3.

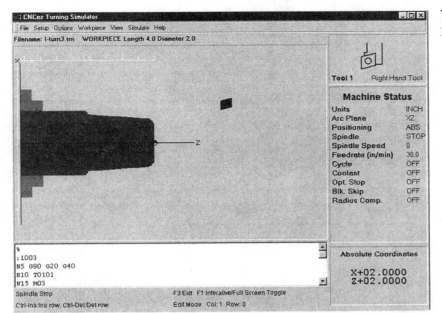

FIGURE 6.55
Interactive Partial Line Editor.

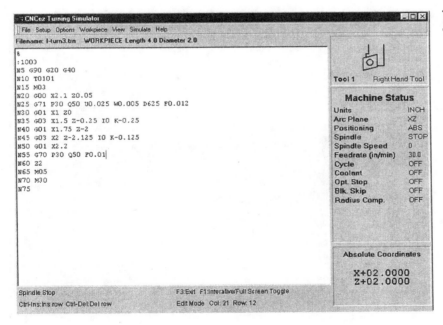

FIGURE 6.56
The full-screen editor.

STEP 7: Go to the Simulate menu and run the complete program, using the Cycle Start option.

From the menu bar, select Simulate.

From the Simulate menu, select Cycle Start.

You will now be able to observe the program simulation and how the G71 and G70 commands take care of many lines of programming (Fig. 6.57).

STEP 8: End of program (Fig. 6.58)

STEP 9: You can optionally save the part program to your working folder.

FIGURE 6.57
G71 command being executed.

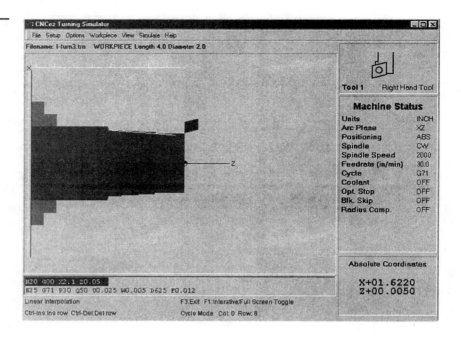

FIGURE 6.58
The completed part.

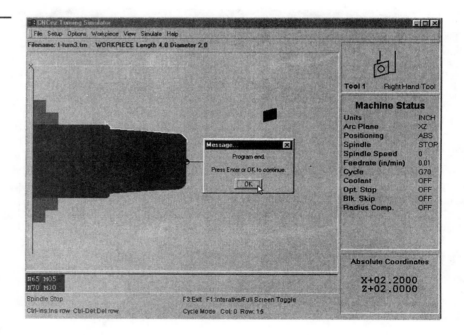

EXAMPLE 4:   I-turn4.trn

This example demonstrates the G72 facing cycle and the G74 drilling cycle commands. The completed part is shown in Fig. 6.59.

Workpiece Size: 2" Diameter by 3" Length

Tool: Tool #2, Right-hand Turning Tool

Tool #3, 3/8" Drill

Tool Start Position: X2, Z3

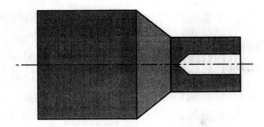

FIGURE 6.59
The completed part.

```
%
:1004
N5 G90 G20 G40
N10 T0202
N15 M03
N20 M08
N25 G00 X2 Z0.05
N30 G72 P35 Q50 U0.05 W0.005 D500 F0.012
N35 G01 X1 Z0.05
N40 Z-1
N45 X2 Z-1.5
N50 X2.2
N55 G70 P35 Q50 F0.006
N60 T0200 G00 X4 Z3
N65 T0303
N70 G00 X0 Z0.1
N75 G74 Z-1 F0.05 D0 K0.125
N80 G00 X4 Z3 M09
N85 T0300 M05
N90 M30
```

STEP 1: Create a new file called I-turn4.

Move the pointer to the menu bar and select File.

Select New...

Enter the file name: I-turn4 [Enter] or Click on OK.

STEP 2: Set up the workpiece (stock material) for this program.

From the menu bar, select Workpiece.

Select New...

Enter the length (in.): 3.

Enter the diameter (in.): 2.

Click on OK or [Enter].

STEP 3: Begin entering the program and simulate the cutter path. Use the Simulate/Edit option.

STEP 4: Program setup phase. You must enter all the setup parameters before you can enter actual cutting moves.

>% [Enter]   Program start flag
>:1004 [Enter]   Program number 1004
>N5 G90 G20 G40 [Enter]   Absolute and inch programming
>N10 T0202 [Enter]   Tool change to Tool #2
>N15 M03 [Enter]   Spindle on clockwise
>N20 M08 [Enter]   Coolant pump 1 on

STEP 5:   Material removal phase.

>N25 G00 X2 Z0.05 [Enter]   Fig. 6.60

FIGURE 6.60
Home position for G72.

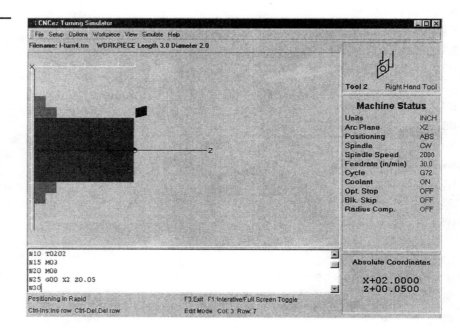

>N30 G00 X1 Z0.05 [Enter]   To be replaced later with G72
>N35 G01 X1 Z0.05 [Enter]   Beginning of profile
>N40 Z-1 [Enter]   Feed in to Z-1
>N45 X2 Z-1.5 [Enter]   Diagonal feed to (X2, Z-1.5)
>N50 X2.2 [Enter]   End of profile (Fig. 6.61)
>N55 G00 Z0.05 [Enter]   To be replaced by G70
>N60 T0200 G00 X4 Z2 [Enter]   Fig. 6.62

STEP 6:   Use the File/Edit option to insert the G72 and G70
              commands.

              To exit the current mode, press F3.

              From the menu bar, select File.

              From the File menu, select Edit.

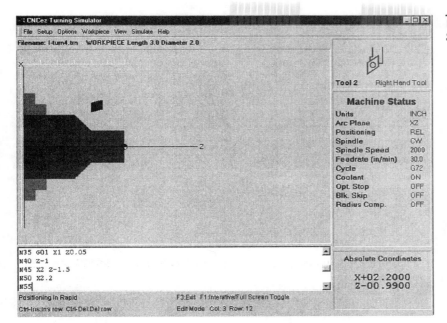

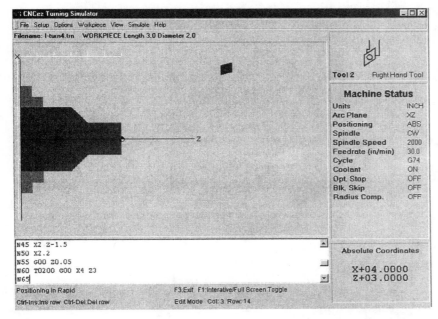

Here you must replace N30 and N55 with the following, also shown in Fig. 6.63.

>N30 G72 P35 Q50 U0.05 W0.005 D500 F0.012   Facing cycle
>N55 G70 P35 Q50 F0.006   Finishing cycle

STEP 7:   Go to Simulate/Edit to complete the program.
         To exit from the editor mode, press F3.

FIGURE 6.63
Replacing N30 and N55 with G72
and G70.

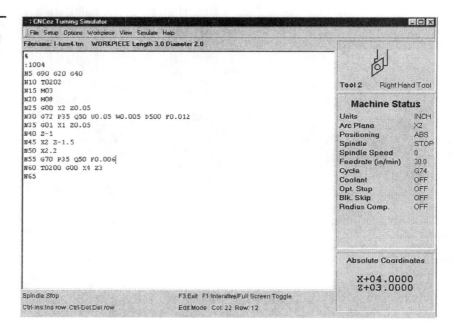

From the menu bar, select Simulate.

From the Simulate menu, select Edit.

You will now be able to observe the program simulation and how the G72 and G70 commands take care of many lines of programming (Figs. 6.64 and 6.65).

STEP 8: Enter the drill cycle command.

FIGURE 6.64
The program running the G72
command.

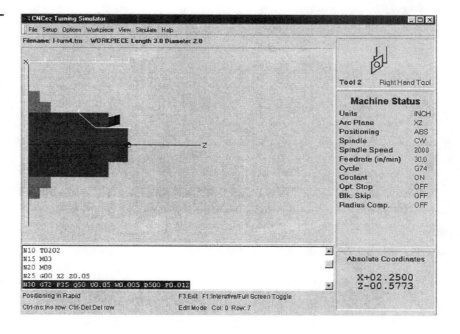

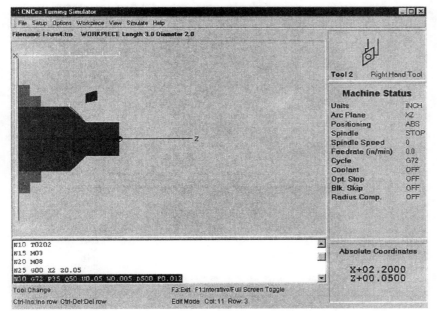

FIGURE 6.65
The program running the G70 command.

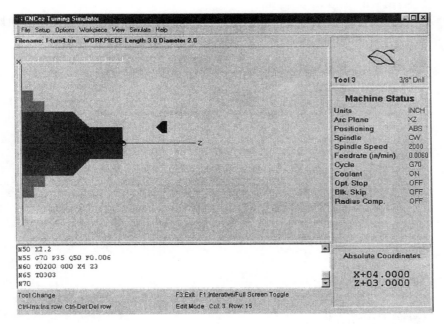

FIGURE 6.66
Calling for a tool change to a 3/8 in. drill.

>N65 T0303 [Enter]   Fig. 6.66

>N70 G00 X0 Z0.1 [Enter]   Rapid move to (X0, Z0.1)

>N75 G74 Z-1 F0.05 D0 K0.125 [Enter]   Fig. 6.67

STEP 9:   Program shutdown phase

>N80 G00 X4 Z3 M09 [Enter]   Rapid move to (X4, Z3) and coolant off

FIGURE 6.67

After the G74 command.

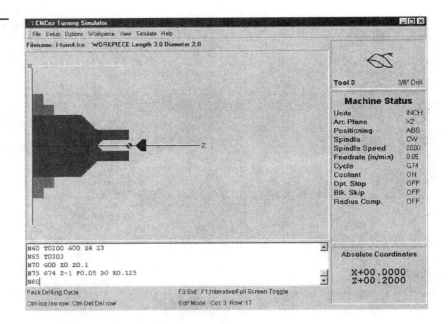

>N85 T0300 M05 [Enter] Spindle off

>N90 M30 [Enter] Fig. 6.68

STEP 10: End of program

STEP 11: You can optionally save the part program to your working folder.

FIGURE 6.68

Program end.

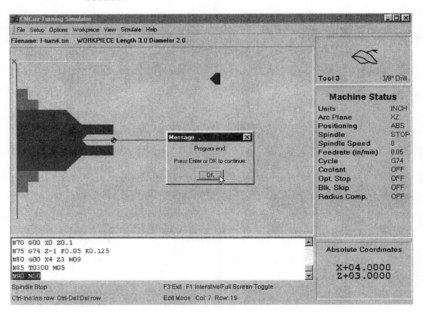

EXAMPLE 5: I-turn5.trn

Workpiece Size: 2" Diameter by 4" Length

Tool: Tool #1, Right-hand Turning Tool

Tool #2, Right-hand Finishing Tool

Tool #3, Grooving Tool

Tool #5, Neutral Tool

Tool Start Position: X2,Z3

```
%
:1005
N5 G90 G20 G40
N10 T0101
N15 M03
N20 G00 X2.05 Z0.05 M07
N25 G71 P30 Q40 U0.05 W0.05 D500 F0.012
N30 G01 X1.5 Z0.05
N35 Z-3
N40 X2.05
N45 T0100 G00 X4 Z3
N50 T0202
N55 G00 X2.05 Z0.05
N60 G70 P30 Q40 F0.006
N65 T0200 G00 X4 Z3
N70 T0505
N75 G00 X2.25 Z-3
N80 G75 X1.25 Z-0.25 F0.25 D0 I0.125 K0.125
N85 T0500 G00 X4 Z3
N90 T0303
N95 G00 X1.5 Z0.05
N100 G01 Z0 F0.012
N105 G76 X1.5 Z-2.75 D625 K0.125 A55 F0.1
N110 G00 X4 Z3 M09
N115 T0300 M05
N120 M02
```

The completed part is shown in Fig. 6.69.

FIGURE 6.69
The completed part.

STEP 1: Create a new file called I-turn5.

Move the pointer to the menu bar and select File.

Select New...

Enter the file name: I-turn5 [Enter] or Click on OK.

STEP 2: Set up the workpiece (stock material) for this program.

From the menu bar, select Workpiece.

Select New...

Enter the length (in): 4.

Enter the diameter (in): 2.

Click on OK or [Enter].

STEP 3: Begin entering the program and simulate the cutter path. Use the Simulate/Edit option.

From the menu bar, select Simulate.

Select Edit.

STEP 4: Program setup phase. You must enter all the setup parameters before you can enter actual cutting moves.

>% [Enter]	Program start flag
>:1005 [Enter]	Program number 1005
>:N05 G90 G20 G40 [Enter]	Absolute and inch programming
>:N10 T0101 [Enter]	Tool change to Tool #1
>N15 M03 [Enter]	Spindle on clockwise

STEP 5: Material removal phase

>N20 G00 X2.05 Z0.05 M07 [Enter]	Rapid move to (X2.05, Z0.05)
>N25 G00 X2.05 Z0.05 [Enter]	To be replaced with G71
>N30 G01 X1.5 Z0.05 [Enter]	Feed move to (X1.5, Z0.05)
>N35 Z-3 [Enter]	Feed move to Z-3
>N40 X2.05 [Enter]	Feed move to X2.05
>N45 T0100 G00 X4 Z3 [Enter]	Rapid to (X4, Z3) and Tool #1 cancel
>N50 T0202 [Enter]	Tool change to Tool #2
>N55 G00 X2.05 Z0.05 [Enter]	Rapid to (X2.05, Z0.05)

STEP 6: Insert the G71 and G70 commands.

Here you must replace lines N25 and N60.

Use the arrow keys to move the cursor to line N25 and replace it with the following:

>N25 G71 P30 Q40 U0.05 W0.05 D500 F0.012 [Enter]	Turning cycle (Fig. 6.70)

FIGURE 6.70
Executing the G71 command.

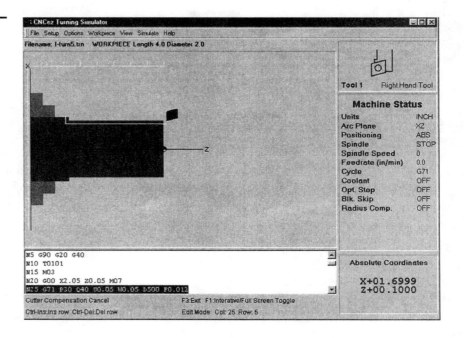

Move the cursor back to the bottom and insert the following:

>N60 G70 P30 Q40 F0.006   Finishing cycle

STEP 7:   Enter the G75 command (Fig. 6.71).

>N65 T0200 G00 X4 Z3 [Enter]   Rapid to tool change position

>N70 T0505 [Enter]   Tool change to Tool #5

>N75 G00 X2.25 Z-3   Rapid to (X2.25, Z-3)

>N80 G75 X1.5 Z-0.25 F0.25 D0 I0.125 K0.125 [Enter]   Grooving cycle

STEP 8:   Enter the G76 command (Fig. 6.72).

>N85 T0500 G00 X4 Z3 [Enter]   Rapid to X4,Z3 and Tool #5 cancel

>N90 T0303 [Enter]   Tool change to Tool #3

>N95 G00 X1.5 Z0.05 [Enter]   Rapid to (X1.5, Z0.05)

>N100 G01 Z0 F0.012 [Enter]   Feed move to Z0

>N105 G76 X1.5 Z-2.75 D625 K0.125 A55 F0.1 [Enter]   Threading cycle (Fig. 6.72)

>N110 G00 X4 Z3 M09 [Enter]   Rapid to (X4, Z3) and coolant off

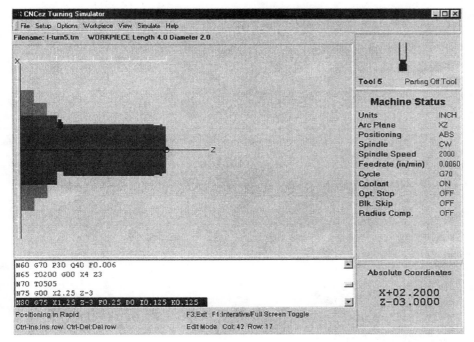

FIGURE 6.71
Grooving cycle.

STEP 9: Program shutdown phase

>N115 T0300 M05 [Enter]   Spindle off and cancel tool

>N120 M02 [Enter]   End of program (Fig. 6.73)

STEP 10:   End of program

STEP 11:   You can optionally save your part program to your working folder.

FIGURE 6.72
Threading cycle in progress.

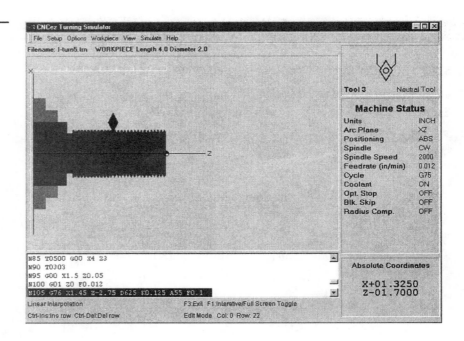

FIGURE 6.73
End of program dialog prompt.

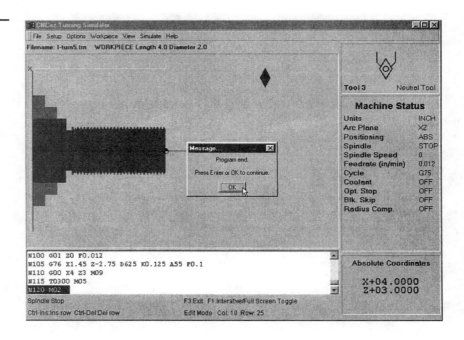

LAB EXERCISES

1. What does the preparatory function G00 command do?

2. How is tool nose radius compensation called?

3. Give an example of a linear feed move.

4. What does the address U stand for when a G71 command is programmed?

5. What does a G76 command specify?

6. Which G-code and additional letter address are used to call up a dwell cycle?

7. Write an example start line for a G74 peck drilling cycle.

8. Which M-code is used to specify spindle on clockwise?

CHAPTER 7

Introduction to CAD/CAM

CHAPTER OBJECTIVES

After studying this chapter, the student should have knowledge of the following:

Basics of computer-aided design

Basics of computer-aided manufacturing

Introduction to AutoCAD

Introduction and tutorial on EdgeCAM

Introduction and tutorial on MasterCAM

THE BASIC CAD/CAM SYSTEM

Originally, computer-aided design–computer-aided manufacturing (CAD/CAM) systems comprised mainly expensive mainframe computers and minicomputers. Because of advancements in computer technology, the personal computer (PC) is now the computer of choice for most users. CAD/CAM software can be general in design for use in all disciplines, or it can be specialized with a specific design goal in mind. A good example of a specialized system is one used primarily for circuit board design. An excellent choice for an all-round CAD software is AutoCAD by Autodesk, Inc. AutoCAD is an open architecture CAD software, which means that it can be run with virtually thousands of third-party add-on packages for a user's specific application.

A typical modern PC-based CAD system may consist of the following:

- *An IBM or compatible personal computer.* The minimum system requirements for AutoCAD R14 as recommended by Autodesk, Inc. are:

   A 90 MHz or better Pentium-based machine

   A minimum of 32 MB of RAM

   A minimum of 100 MB of free hard disk space

   4X or better CD-ROM drive

   Windows 95/98, NT 4.0 or greater

- *A Windows-supported video display adapter.* The minimum is generally an SVGA standard that supports 800 x 600 pixels. Many types of video adapters are available from various manufacturers, including numerous CAD accelerator cards with support for OpenGL.

- *An input device.* Some input devices that can be used include a Microsoft or compatible mouse and/or a digitizing tablet such as a Summagraphics Summasketch, Hitachi, or Kurta Tablets.

- *An output device.* Some output devices supported by AutoCAD include various printers and plotters, such as the Hewlett-Packard Draftpro and the Calcomp Drawingmaster. Any Windows-supported printer (e.g., laser printer) can be used as an output device.

COMPUTER-AIDED DESIGN

At its most basic, CAD is a geometric modeling system used to produce two- or three-dimensional engineering drawings of parts. Although the term is sometimes used to describe computer-aided drafting, CAD involves the use of a computer in the total design process.

Design is a process that involves identifying a need, generating possible solutions to meet that need, evaluating each solution to determine its merits (engineering analysis), identifying the solution to be developed based on merit, and developing a detailed model from which a part can be built.

The computer can aid in most steps in the design process. The extent to which it can do so depends on the cost and availability of computer hardware and software. Over the past decade, costs have dropped to the point where a CAD system is very affordable, even for the smallest of companies. This is still the trend, so more and more functions are being added to the existing power of CAD. These functions include finite element analysis (FEA), finite element modeling (FEM), parametric design, and three-dimensional modeling. Currently, one of the primary uses of CAD is in the production of three-dimensional geometric models from which engineering drawings and CNC part programs are produced almost automatically.

The engineering or shop drawing, once considered the culmination of all design efforts, is now mostly considered just a reference document to assist in manufacturing quality control. Even this function may someday disappear as corporations go paperless. A drawing would be developed in the CAD process without the need to fabricate a prototype, so all the testing and analysis could be performed and a part program generated. Once the part has been fabricated, it can then go through a computer-assisted quality control check. All these tasks could be performed without ever having to pick up a shop drawing.

The major goals of CAD for manufacturing are:

- To increase productivity.
- To create a database for manufacturing.

CAD helps the designer (or draftsperson) visualize a design on the computer screen. The designer can make a change and get almost immediate feedback on the results. Current CAD software also allows for analysis and testing of components before manufacture and the presentation of the finished product.

A good CAD system has the following characteristics:

1. *It should be easy to learn.* It should be a step-by-step process and should be done in a logical manner. It should be menu-driven and highly graphic with as much data as possible displayed on screen. It also should have an online tutorial (so that you can get started quickly), online context-sensitive help accessible from any point in the program, and online documentation.
2. *It should be easy to work with.* It should be adaptable to your particular application and not require a specialist to operate it.
3. *It should have macro commands.* A macro is a single key or command sequence that causes the execution of a string of commands.
4. *It should have its own embedded programming language.* It should have its own customization language.
5. *It should be expandable.* It should be able to grow with you. You should also easily be able to change peripheral devices such as display monitors and plotters.
6. *It should have an open database.* It should do more than just create drawings. It should be able to generate bills of material and hold attributes such as price and other database information.
7. *It should be compatible with other CAD/CAM and analysis programs.* It should easily be able to exchange geometric models with other programs, using standard model interchange file formats

such as IGES, DXF, and STL. IGES is the current standard for three-dimensional geometric model exchange involving parametric surfaces (for example, NURBS). DXF (Autodesk) is the de facto standard for PC geometric model exchange. STL (stereolithography) is the de facto standard for rapid prototyping system models (three-dimensional polygon models only). Most analysis programs (for example, finite element, boundary element, computational fluids, kinematics, and so on) accept one or more of these file formats to import a geometric model for analysis.

AUTOCAD

AutoCAD is used by more than two million customers in 85 countries, making it the world's largest-selling PC-based CAD package. Using Auto-CAD, you can create a three-dimensional model of a design and then automatically create two-dimensional working drawings from that model. For presentation purposes, these three-dimensional models can be used again to create true three-dimensional renderings for both visualization and presentation purposes. AutoCAD is used in all fields, from mechanical engineering, architecture, landscape design, and civil engineering to cartography and interior design.

AutoCAD uses an intuitive graphic interface that allows you to execute commands via pull-down menus, icon menus, dialog boxes, the command line, or a digitizer template. Using a mouse or digitizer, you can select tools from the easy-to-use menus to create geometric shapes of any complexity quickly. You can then edit and change these shapes quickly with commands such as mirror, trim, copy, stretch, fillet, chamfer, erase, rotate, move, and scale. Zoom and pan commands let you view the entire drawing on the screen or zoom in to an area no matter how small. Auto-CAD also provides full-dimension control according to ISO, DIN, ANSI, and other standards. With AutoCAD's text tools, you can annotate your drawing easily. After you complete your drawing, AutoCAD's plotting tools let you plot your drawings exactly as they appear on screen. You can use up to 255 combinations of line types, colors, widths, and pen speeds for the same plotter.

AutoCAD also uses an open architecture, which means that you can customize AutoCAD to suit your specific requirements. There are more than 1000 off-the-shelf, third-party packages that customize AutoCAD, from which you can choose to add to your system. These programs integrate with AutoCAD to create specialized systems for architectural design, mechanical drafting, structural engineering, mapping, mining, landscape design, fashion design, circuit board design, and dozens of other applications. With new Windows object linking and embedding (OLE), as well as visual basic for applications (VBA), programming interfaces allow third-party developers of applications to integrate directly into the AutoCAD application. An example is the EdgeCAM machining solution described in the next section.

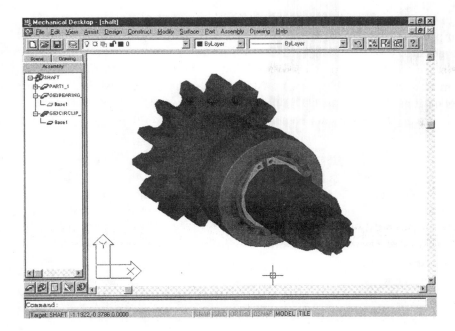

A screen display of Autodesk's Mechanical Desktop.

AutoCAD's ability to adapt to and work with specific programs helps make AutoCAD even more useful. AutoCAD drawings can also be transferred to other CAD or CAM packages, such as EdgeCAM and Master-CAM.

The major benefits of a CAD package like AutoCAD are that it:

- Automates repetitive tasks
- Reduces duplicate efforts
- Allows you to communicate with others electronically
- Makes drawing revisions easier to perform
- Enables you to customize your environment and tailor AutoCAD to meet your needs
- Lets you create extremely accurate and high-quality drawings
- Provides impressive presentations
- Supports a wide selection of hardware

COMPUTER-AIDED MANUFACTURING

Computer-aided manufacturing (CAM) utilizes computers to support and control manufacturing operations. There are two main applications for CAM:

1. Those whereby the computer directly controls a manufacturing operation.
2. Those whereby the computer is used to support manufacturing operations—for example, in inventory control and CNC part programming.

In the past, CAM, as it applies to NC part programming, also was called computer-assisted programming. Generally, CAM involves either defining the geometry of a part or calling it up from an existing file. Then it describes the cutter toolpath including feeds, speeds, direction, coolant, and clamping. Finally, it generates a CNC file through a process called postprocessing.

There are three distinct steps in the CAM process:

1. *Input or define the part geometry.* Most CAM systems enable the programmer to describe the part geometry. They also allow the importing of data from other software packages such as CAD. Today's CAD systems actually evolved from early computer-assisted programming and computer graphics programming systems.

2. *Describe the cutter toolpath.* Describing the toolpath involves selecting the tools to be used for a particular job, specifying the feeds and speeds, and activating the clamps and coolant. Each of the many CAM systems on the market today works slightly differently. The systems that are more difficult to use involve use of the centerline cutter toolpath. The part program is modified by erasing certain geometry or by changing the tool information to create offsets and pockets. The packages that are better and easier to use employ the part geometry as a reference, and the programmer describes the toolpath with it.

3. *Generate the final CNC program.* The final CNC program is generated when the postprocessor is run. In many older CAM systems, the postprocessor was a separate software package. Today's systems have integrated sophisticated postprocessor programs. Many

A screen display of a drafted part using EdgeCAM V3.5.

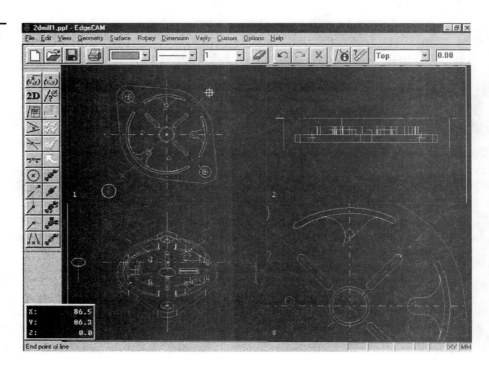

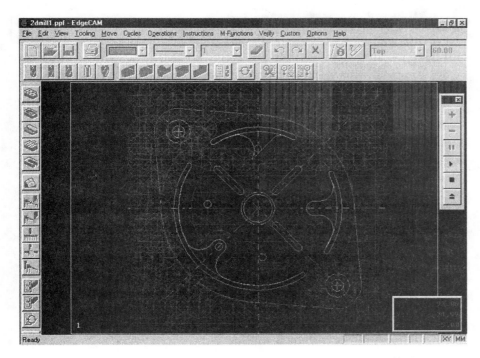

Modern CAM packages such as EdgeCAM automatically calculate and generate toolpaths for area clearance in pocket milling operations.

companies develop postprocessors for other companies. A postprocessor, whether integrated or separate, is customized for a particular machine tool. As discussed in Chapter 1, there are many variations of machine tool controllers and a different postprocessor is needed for each.

CAM software packages such as EdgeCAM and MasterCAM help provide the CNC programmer with a valuable productivity tool for both the generation of CNC part programs and process planning. They help reduce the time it takes to generate accurate machine-ready NC programs. Geometric data from CAD software such as AutoCAD are easily translated into these programs, which can be easily downloaded to a CNC machine controller directly from the computer. Today machine shops actually have dedicated PCs running CAM software packages right on the shop floor.

EDGECAM

EdgeCAM covers a wide range of machining operations, including milling, turning, and wire erosion. A modular CAM package, EdgeCAM starts with simple 2½-axis milling or 2-axis turning and moves on to cover surface and rotary milling and 4-axis turning. EdgeCAM's flexibility can meet many manufacturing requirements. EdgeCAM is designed to be intuitive and simple to understand, minimizing training requirements and learning curves. It has been designed specifically for the familiar Windows 95 and NT operating systems, so it makes best use of these environments. Dropdown menus, icons, tool tips, dialog boxes, and wizards are just some methods of communicating with the system that you will recognize from using other Windows products such as word processing and spreadsheet programs.

As an application downstream from CAD, EdgeCAM performs an important communication function between products. EdgeCAM is extremely versatile in reading data from other systems. It supports DWG/DXF, Microstation, IGES, SAT, and VDA-FS formats. EdgeCAM for Mechanical Desktop is fully integrated with Autodesk's Mechanical Desktop. It directly machines solid models designed with Autodesk's parametric feature-based solid and surface modeling functions. It also automates the NC programming process inside AutoCAD R14 (Fig. 7.1). This is extremely beneficial to the designer because any changes to the solid model can be immediately seen in the CNC part program generated by EdgeCAM.

Being a comprehensive CNC programming solution, EdgeCAM supports vital functions to maximize machine tool productivity such as:

- 2-D, 3-D, and surface design
- 2½ to 3-axis milling and surface machining
- Rotary, A-, B-, and C-axis support
- Multiplane milling
- 5-axis machining, trimming, and deflashing
- 2- and 4-axis and live tooling; C- and Y-multiaxis turning
- 2- and 4-axis wire EDM
- Tool radius compensation

FIGURE 7.1
EdgeCAM can run inside Autodesk's Mechanical Desktop.

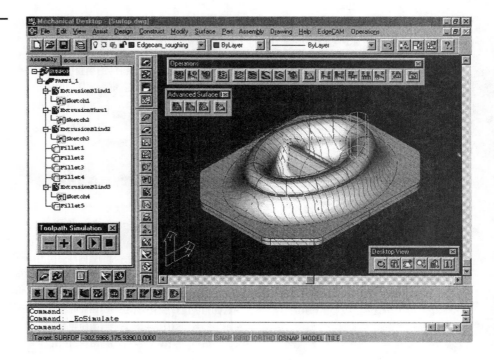

- Canned cycles
- Geometry creation importing and editing
- Profiling, slotting, and face milling tasks
- Hole pattern drilling
- Unique thread milling features
- Interactive solid verification displays of tools, toolpath, clamps, and fixture as solids
- Family of parts machining
- Keyboard macros and custom programming capabilities
- Common database, allowing multiple machining from a single file
- RS-232 communication support to machine tools
- Tool Library facilities
- Advanced CNC code editor
- Open architecture with C++ links for developing custom routines
- Sophisticated code wizard assists in creating custom postprocessors
- Seamless integration with Autodesk Mechanical Desktop

These advanced features set EdgeCAM apart from the average Windows-based CAM solution.

The following introduces the basic functions of EdgeCAM, including the Design module and manufacturing features.

INTRODUCTION TO EDGECAM

EdgeCAM allows interactive graphic creation and modification of CNC programs using the simplest and best-known machine-based user interface.

EdgeCAM provides full:

2D/3D Design and Modeling

$2\frac{1}{2}$- to 3-axis machining

Multiaxis turning

To run EdgeCAM from your Windows desktop, select the Start button from the task bar. Select the Programs option; then select the EdgeCAM group. Now select the EdgeCAM item that loads EdgeCAM.

The EdgeCAM screen is similar to other Windows applications. It contains several elements that will be familiar.

EdgeCAM Modes

EdgeCAM can operate in one of two modes: Design and Manufacture. By default EdgeCAM is opened in Design mode. This is the mode in which you can design or edit the part designs. The Manufacture mode is used to create and edit toolpaths. There are several manufacturing disciplines that can be selected from the Manufacture mode, including Milling and Turning.

COORDINATE SYSTEMS IN EDGECAM

EdgeCAM can accept coordinate data using a variety of coordinate systems such as:

Cartesian

Polar

Angular

You can select the type of coordinate system to be used in the Coordinate Input dialog box which is activated by the coordinate button. This is the top right-hand button in the Command Modifier Toolbar (Fig. 7.2).

SELECTING THE ENVIRONMENT

EdgeCAM operates within an environment that is selectable according to the type of part being designed or manufactured.

There are two environments to choose from:

XY Environment

This provides view ports and construction planes for constructing a part with respect to the usual X, Y, Z Cartesian coordinate system. This is the default environment when entering EdgeCAM. Typically, the XY environment is used for generating Milling.

ZX Environment

This environment provides view ports and construction planes that simulate the orientation of a turned part in a machine tool during manufacture. In the case of a lathe, for example, it is usual to work in the ZX plane where the Z axis is horizontal.

FIGURE 7.2
The command modifier toolbar is a useful aid in entering coordinate data.

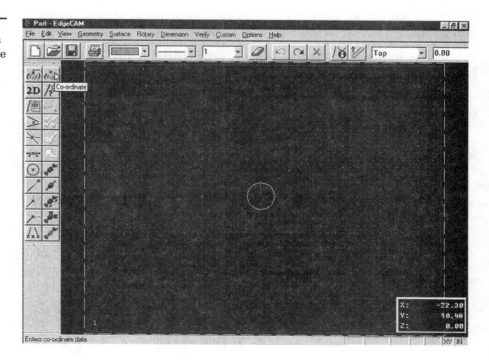

Using Construction Planes (CPLs)

The EdgeCAM database maintains part models with respect to a coordinate system known as the World coordinate system. To help you construct a model, the concept of the construction plane (or CPL) is used. A construction plane defines a local coordinate system at any orientation to the World axes. As each CPL has its own x, y, z-axes, you only have to deal with local coordinates when creating entities. EdgeCAM translates CPL coordinates into World coordinates automatically for you.

Predefined CPLs

A set of predefined CPLs is provided for the XY and ZX environments (Fig. 7.3). The Construction Planes are as follows:

Top

Back

Left

Front

Right

Bottom

If you need to add a feature, such as a slot, into the bottom of the part, you can select the "bottom" CPL and work in terms of the dimensions you have. Don't worry about how they map back to the World axes.

The Drawing CPL

The Drawing CPL is different from all the other CPLs as it is purely two dimensional. It is like a conventional drawing sheet on which views are

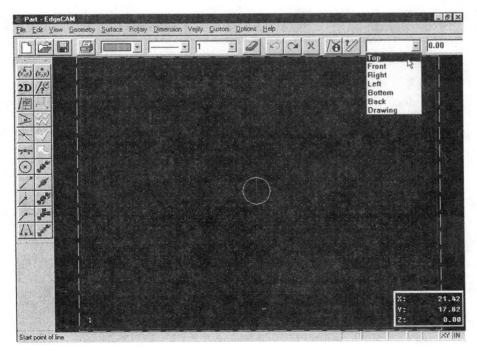

FIGURE 7.3
EdgeCAM's predefined CPLs.

assembled to provide the views that comprise an engineering drawing. Its primary purpose is to contain elements of a drawing that are not part of the 3-D model, such as dimensions, text, and drawing symbols. By making the Drawing CPL current, it can also be used for making two-dimensional drawings.

Entity Types

EdgeCAM provides several entity types that you can use when designing and manufacturing components:

Point entities
Line entities
Circle and arc entities
Curve entities
Surface entities
Continuous entities
Group entities
Detailing entities
Toolpath entities

Using Metric and Imperial Units

You can work in EdgeCAM using either metric or inch units. The units are one of the parameters that can be set with the System (Options menu) option. The selected units are saved as part of your personal defaults and automatically used until reset.

Note that many areas of EdgeCAM can work simultaneously with different unit types. For example, the part units could be metric, but you are generating toolpaths with an inch tool.

CREATING AND EDITING GEOMETRY

There are several stages involved when creating geometric entities. The typical stages are:

1. Select the command for creating the required entity type. This is usually a command from the Menu bar. Toolbar buttons are provided for frequently used commands.
2. Enter parameters into a dialog box to modify the effects of the command. This stage is usually bypassed by command buttons.
3. Enter coordinate data to position the entity, either directly on screen by clicking the mouse ("digitizing") or by entering explicit coordinates in a dialog box.

Here is a typical example of creating a geometric entity:

1. Select Geometry from the Menu bar. A pop-up menu appears.
2. Select the required entity type. A dialog box shows the parameters available for a command.
3. Enter the appropriate values for the parameters.
4. Click the OK button at the bottom of the dialog box to accept the values.

5. You may now have to digitize various points in the drawing frame depending on the type of command you have selected. Look at the prompt in the Status bar for instructions.
6. Select the Finish button to complete various parts of the command or to complete the command. This is the Green Check Mark button of the Command Modifier toolbar.

SELECTING COMMAND BUTTONS ON TOOLBARS

The command buttons on the various toolbars provide a subset of commonly used commands, such as creating a line through two points. Because they are set up to perform specific functions, it is quicker to use them rather than to choose their equivalents from the Geometry menu. Prompts for the required data will appear in the Status bar on the bottom left of the screen. Select the Finish button to complete various parts of the command or to complete the command.

To display toolbars:

1. Select the View, Toolbar menu item.
2. Select the required toolbars in the Select Toolbars dialog box.
3. Click OK.

Entering Coordinate Data Using the Mouse

After you have selected a command, entered parameters, and clicked the OK button to accept these values, you may have to enter coordinate information for the command. Prompts will appear in the Status bar to direct input of the coordinate and entity data required by the command. You can enter this information by moving the cursor around in a viewport of the graphics screen area and clicking the mouse ("digitizing"). The basic methods used for input are coordinate ("free") digitizing, entity digitizing, and grid digitizing.

Coordinate Digitizing (Left Mouse Button)

The basic method for providing coordinate data to a command is to use the mouse to move the cursor to the required location and then click the left mouse button. This provides absolute X, Y coordinates with respect to the current CPL, with the Z ordinate taken from the level of the current CPL. This is also known as "free digitizing."

Entity Digitizing (Right Mouse Button)

Where commands require an entity to be selected, use the mouse to move the cursor onto the entity and press the right mouse button.
Where the coordinates of the end of an entity are required, place the cursor on the required end and press the right mouse button.

Grid Digitizing (Shift + Left Mouse Button)

A grid can be displayed using the Grid (View menu) command. When digitizing, you can snap to a grid point by moving the cursor near to the point, holding down the Shift key, and pressing the left mouse button.

Configuring the Grid

You may find snapping to grid positions useful for positioning entities. To configure the grid, select the View, Grid Configure menu command. On selecting the command, you can specify these parameters:

Display—Mark this box to show the grid in the active view port. You can still snap to a grid point even if the grid is not visible.

X Space—Specify the horizontal space between grid points in part units.

Y Space—Specify the vertical space between grid points in part units.

Methods for Entering Coordinates

The submodifier toolbar provides more advanced facilities for selecting entities or specifying coordinates. You can select entities individually by digitizing each one with a right mouse click. Alternatively, the selection toolbar provides many other methods of selecting entities. Not all of these buttons may be active for a given command. If 2D Snap has been selected, the Z ordinate will be set at the level of the current CPL. Otherwise, the full 3D coordinates of the endpoint are passed to the command.

Once you have selected the entities, select the Finish button.

EXAMPLE OF CREATING GEOMETRY IN EDGECAM

In the following example a quick overview of EdgeCAM's features is covered. You will draw a simple 8 x 4-inch rectangle.

Start by selecting Geometry from the Menu Bar, and choose Rectangle. A Dialog Box will appear on the screen. Since we want an 8 x 4 inch box, we will enter in a length of 8 and a width of 4. Now click on OK (Fig. 7.4).

The system will now ask for the start of the rectangle. The status bar in the bottom left corner of the screen will display this prompt. Choose the

FIGURE 7.4

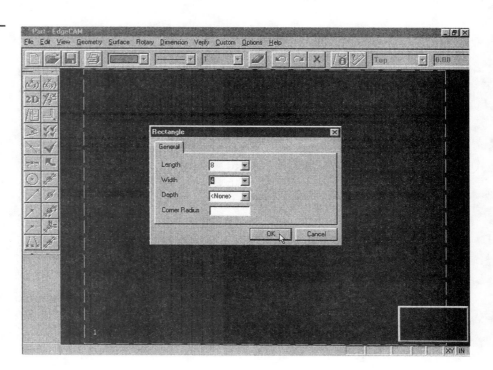

FIGURE 7.5

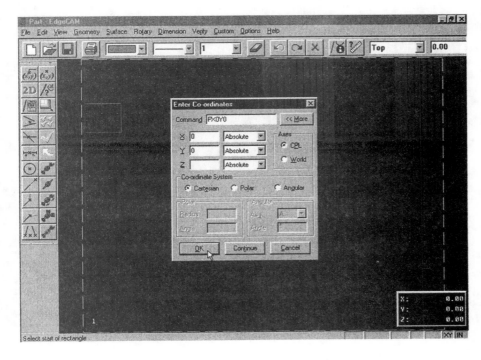

Coordinate Icon from the top right-hand corner of the Command Modifier toolbar.

Enter in X0, Y0 and then click OK (Fig. 7.5).

You will now see a rectangle in the center of the screen. Choose the Green Check Mark from the Command Modifier Toolbar to exit the Rectangle Command.

The size of the drawing displayed may be quite small on the screen, depending on how the system is set up (Fig. 7.6).

FIGURE 7.6

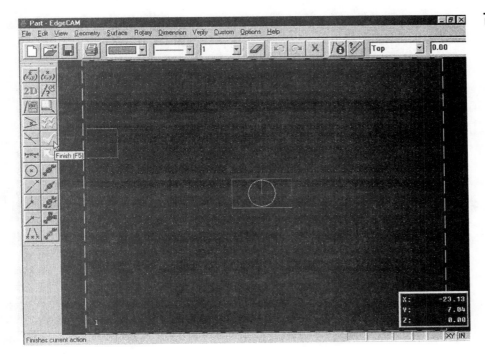

FIGURE 7.7

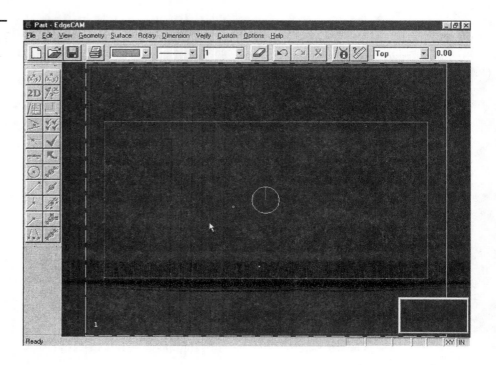

In order to change the size or to fit the drawing to the screen, select View, then Zoom Extents. The drawing will now fit on the screen. The size of the box has not changed, only the way it is displayed on the screen (Fig. 7.7).

You will now draw an 8 x 4-inch stock part using a different technique. You will first have to delete the original drawing. From the Menu Bar select Edit, Delete. When the entities dialog box appears, click OK. You are asked to "DIGITIZE ENTITY TO DELETE," which may be accomplished in more than one way. You can move the cursor, using the mouse, over the top of the line entities that form the rectangle and press the left mouse button. That line will disappear (Fig. 7.8).

FIGURE 7.8

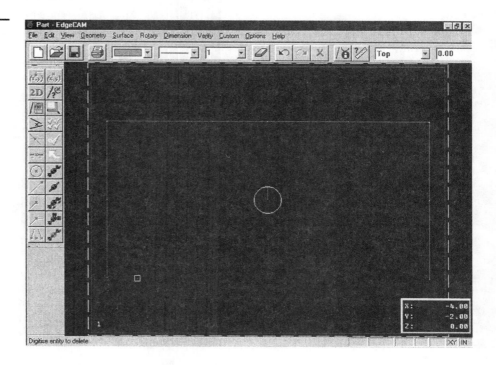

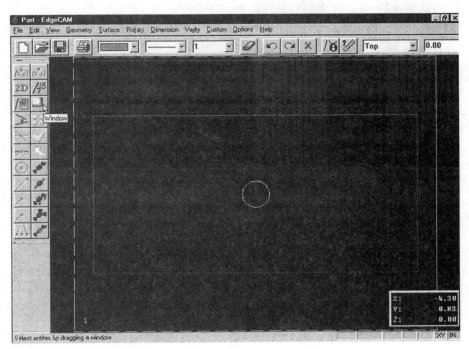

FIGURE 7.9

You could repeat this process until you have deleted all the entities and then choose the Green Check Mark from the Command Modifier Toolbar to exit the Delete Command.

A simpler way would be to choose the Windows Icon from the Command Modifier Toolbar (Fig. 7.9).

A message asking you for the first corner of the window will appear in the lower left corner of the screen. Using the mouse, move the cross hair below the bottom left-hand corner of the rectangle and click the left mouse button. A new message asking for the second corner of the window will appear. This will be the corner that is diagonal to the first corner you picked. Move the mouse to the top right-hand corner of the rectangle making sure that the rectangle is completely inside the window, and press the left mouse button. If the position of the window encompasses the entities to delete, press the right mouse button. Everything inside the window will now change color on the screen, and you will be asked if there are any entities you want to deselect. Since we want to delete all the lines, click the Green Check Mark from the Command Modifier Toolbar (Fig.7.10). Click the Green Check Mark from the Command Modifier Toolbar again to exit the Delete command.

You will now draw an 8 x 4-inch blank stock part, using a different method.

STEP 1:   Create the stock material.

From the Menu Bar select Stock/Fixture from the Geometry pull down menu(Fig. 7.11).

A Dialog Box will appear on the screen. Enter a depth of -1; be sure that the Type is Stock and Shape is Box. The Create Geometry check box should also be selected.

FIGURE 7.10

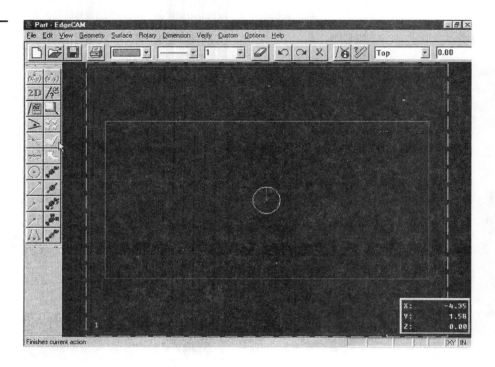

Now click OK.

The system will ask you for the first corner of the box displayed in the status bar on the bottom left corner of the display. Choose the Coordinate Icon from the top right-hand corner of the Command Modifier toolbar.

Enter X0, Y0 and then click Continue (Fig. 7.12).

FIGURE 7.11

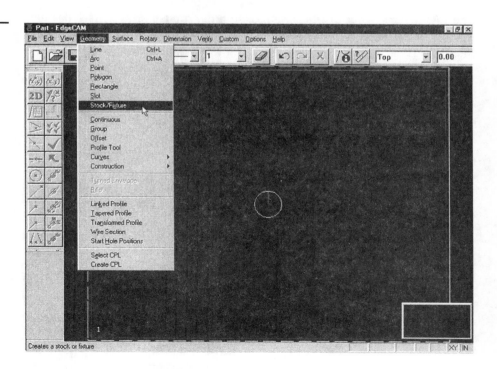

FIGURE 7.12

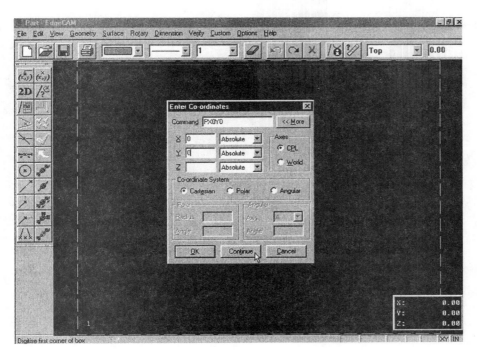

The system will ask you for the second corner of box. This prompt is displayed in the status bar on the bottom left corner of the screen.

In the Enter Coordinates dialog enter an X value of 8 and a Y value of 4, click OK (Fig. 7.13).

FIGURE 7.13

FIGURE 7.14

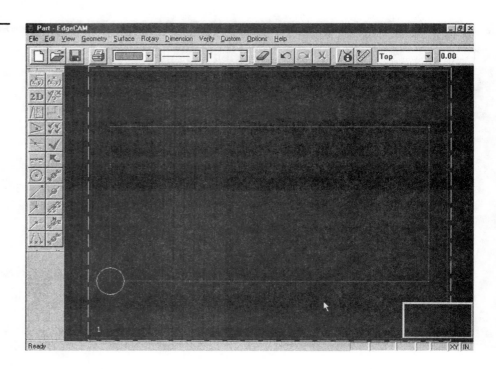

An 8 x 4-inch rectangle will now be displayed (Fig. 7.14). Select View, Zoom, Extents.

STEP 2:   Create a part profile

You will now create the part geometry for this example. First set a better grid setting so it will be easier to enter entity coordinates using grid digitizing.

From the menu bar select View, Grid Configure (Fig. 7.15).

FIGURE 7.15

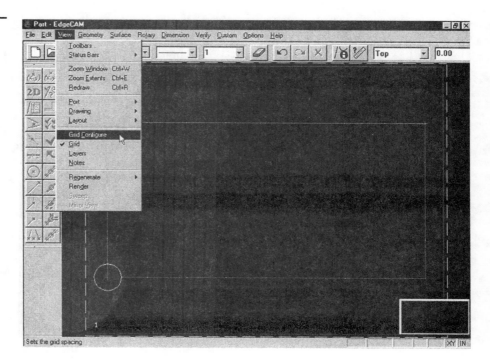

FIGURE 7.16

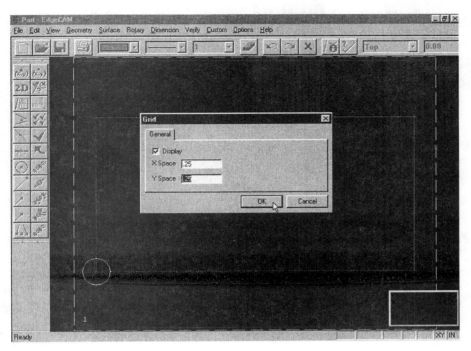

A Grid dialog will be displayed.

Enter an X and Y spacing of .25 and click OK (Fig. 7.16).

A denser grid will now be displayed (Fig. 7.17).

Select the System (Options menu) option and ensure that both the System and Drawing Tolerances are set to .01, and that the track cursor item is checked.

You will now create a profile to be machined later on.

From the menu bar select Geometry, Line (Fig. 7.18).

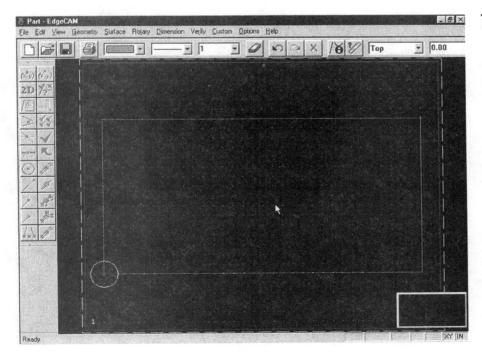

FIGURE 7.17

FIGURE 7.18

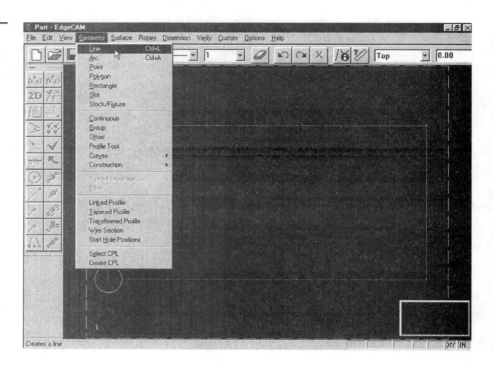

At the Line dialog select the Polyline check box and click OK (Fig. 7.19).

The system will now prompt for a Startpoint for continous geometry creation.

Using the Grid Digitizing method, hold down the Shift key and digitize near X.5, Y.5 using the left mouse button. Digitize another point at X4.5, Y.5 (Fig. 7.20).

FIGURE 7.19

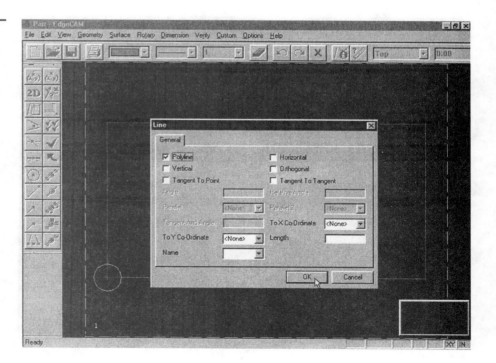

FIGURE 7.20

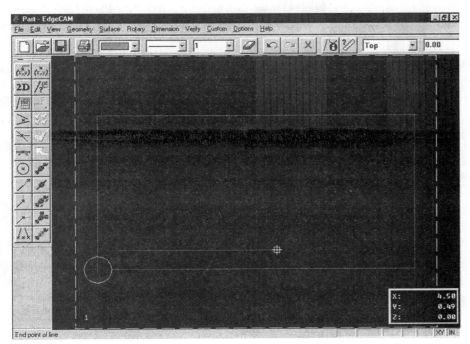

Alternatively you can select the Coordinate Icon from the top right-hand corner of the Command Modifier toolbar and enter the coordinates manually. Click Continue for the rest (Fig. 7.21).

FIGURE 7.21

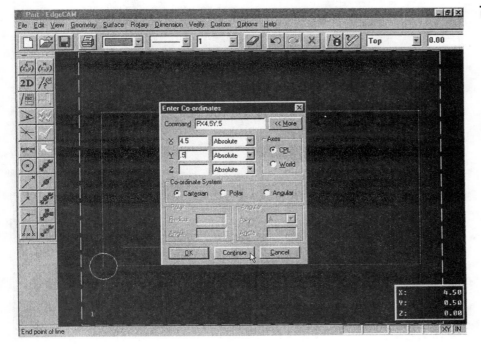

Repeat either process for the following coordinates:

X4.5, Y1.25

X5.75, Y1.25

X7, Y2

X7, Y3.5

X2.75, Y3.5

X2.75, Y2.25

X.5, Y2.25

X.5, Y.5

Now click on the Green Check Mark in the Command Modifier toolbar on the left to complete the command. The screen should look similar to Fig. 7.22.

STEP 3:   Fillet the corners of the profile

You will now dynamically generate fillet radii on several corners of the profile. From the menu bar select Edit, Radius (Fig. 7.23).

At the Edit Radius dialog be sure that Dynamic and Trim are check marked.

Enter a Radius value of .5 (Fig. 7.24).

FIGURE 7.22

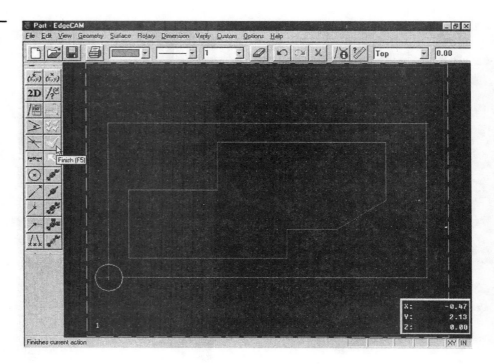

FIGURE 7.23

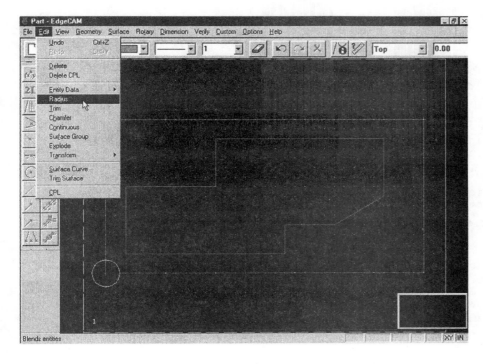

By slowly moving the cursor near a corner, a dynamic radius will be drawn.

Digitize near the bottom left corner using the left mouse button to create the first corner radius (Fig. 7.25).

FIGURE 7.24

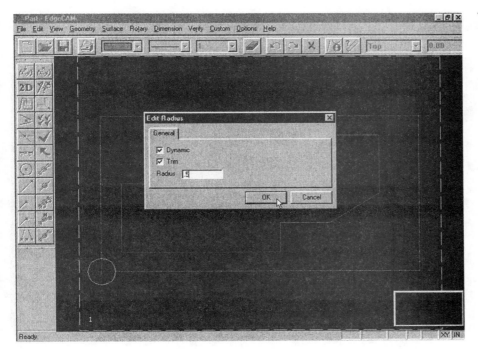

FIGURE 7.25

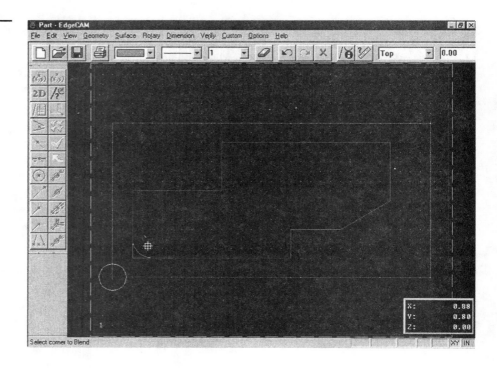

Complete the six corners in a clockwise direction as in Fig. 7.26.

Exit the command by selecting the Green Check Mark in the Command Modifier toolbar on the left.

Now repeat the Edit, Radius command, but this time set a radius value of .25 and create two corner radii as in Fig. 7.27. Remember to complete the command by clicking the Green Check Mark in the Command Modifier toolbar on the left.

FIGURE 7.26

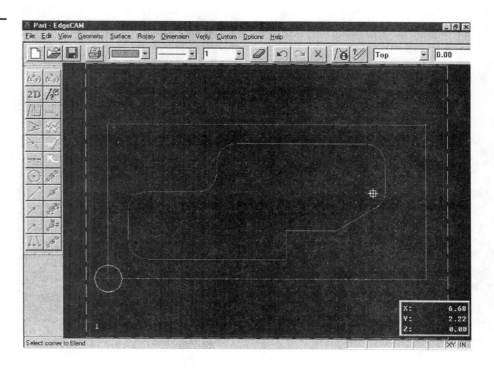

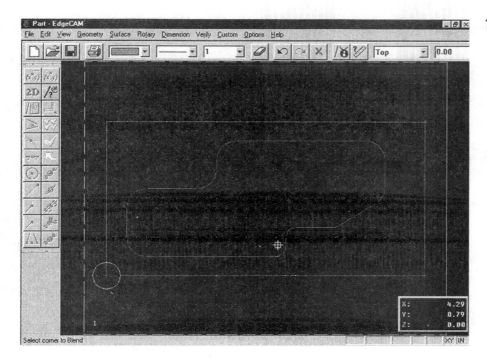

FIGURE 7.27

STEP 4: Save the part geometry

You should now save your geometry. From the file menu select File, Save As.... At the File dialog enter a filename of "geom1" and click Save.

STEP 5: Create the toolpaths

Enter the Manufacturing mode of EdgeCAM.

From the menu bar select the Options, Manufacture menu item (Fig. 7.28).

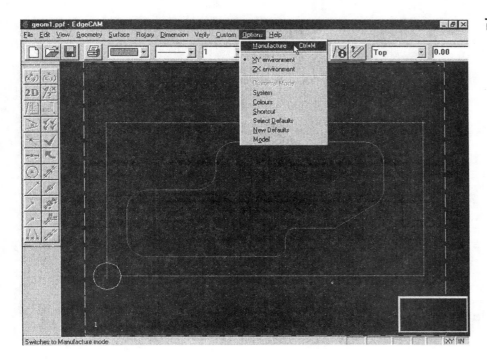

FIGURE 7.28

FIGURE 7.29

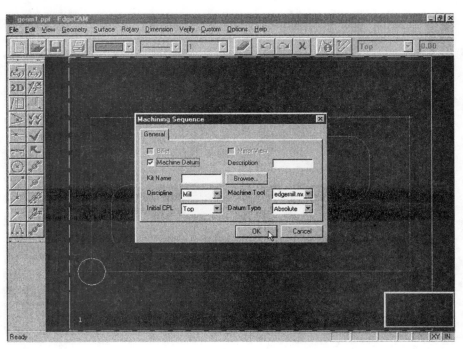

At the dialog prompt be sure that the Machine Datum is checked, Discipline is set to Mill, Initial CPL is set to Top, and click OK. Using the Grid Digitize method, digitize a machine datum point of X0, Y-1. The display will be redrawn and the CPL coordinate box will display X, Y, Z coordinates in red (Fig. 7.29).

From the M-Functions menu select Machine Parameters (Fig. 7.30).

In the Machine Parameters dialog Set an Initial plane of .5, Output Tolerance of .01, and Inch Units. Default the Machine Tool to use the edgemill.mcp file.

FIGURE 7.30

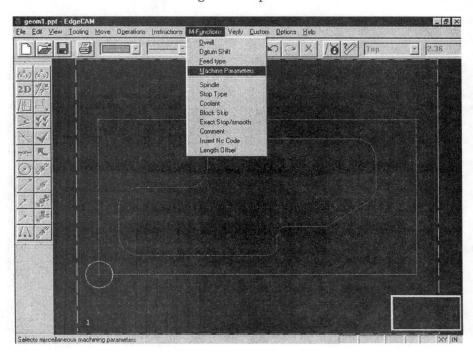

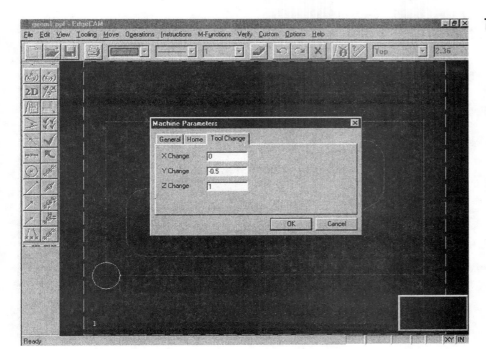

FIGURE 7.31

Now click the Home tab and set the Home position to X0, Y-0.5, Z1. Set the Tool Change settings the same as for the Home Position and click OK (Fig. 7.31).

A shortcut to these and other import commands in the Manufacturing mode can be quickly accessed by right-clicking in the graphics area where a popup menu will be displayed (Fig. 7.32).

You will now create a pocket operation.

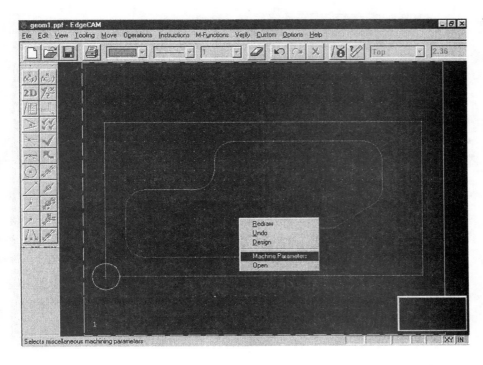

FIGURE 7.32

FIGURE 7.33

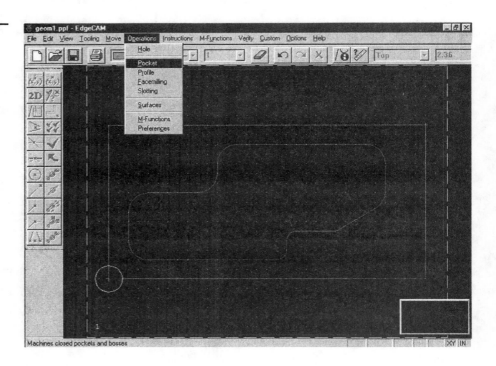

Select the Operations, Pocket menu item from the menu bar (Fig. 7.33).

You will be asked to digitize the pocket profile as shown in the status bar prompt on the bottom left of the screen.

From the Command Modifier toolbar select the chain button (Fig. 7.34).

FIGURE 7.34

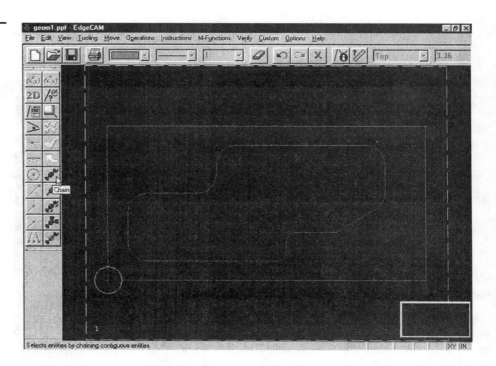

FIGURE 7.35

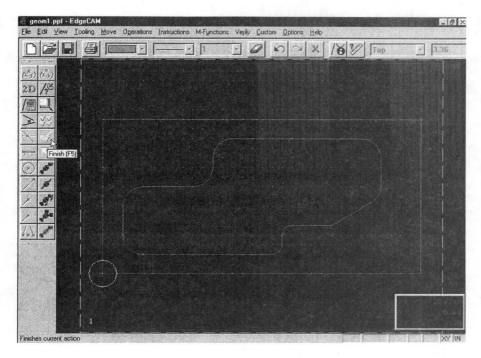

Digitize the bottom horizontal line of the profile. The hole profile will change color and two Xs will appear.

This confirms the chain command was successful. Click the Green Check Mark button of the Command Modifier toolbar to complete this comand (Fig. 7.35).

Click the Check Mark button again since no more profiles exist.

The Pocket Operation dialog will now be displayed.

Set the following values to complete the General pocket operation settings:

Clearance	0.05
Retract	0.125
Level	0
Depth	-0.75
Pocket Type	Blind Pocket
Mill Type	Conventional

Select the Roughing tab and set the following values:

Strategy	Lace
Cut Increment	(Leave blank)
Stepover	40
Offset	0.1
Lace Angle	45
Feedrate	15
Speed	1200

Diameter	0.375
Plunge Feed	5
Position	1

Select the Finishing Tab and set the following settings:

Strategy	Profile
Lead In/Out	None
Toolpath	Centerline
Feedrate	10
Speed	1500
Diameter	0.25
Plunge Feed	5
Position	2

Now Click OK (Fig 7.36).

The pocket and finishing toolpaths will be automatically generated by EdgeCAM (Fig. 7.37).

STEP 6: Generate the CNC code

At this point you can generate CNC code.

From the File menu select Generate Code (Fig. 7.38).

FIGURE 7.36

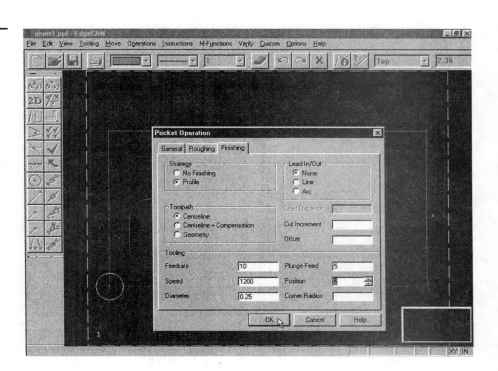

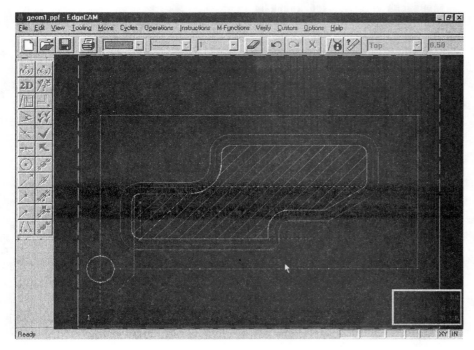

FIGURE 7.37

From the Generate CNC Code dialog, click the Browse button to specify the filename and location of the CNC code file you will be generating.

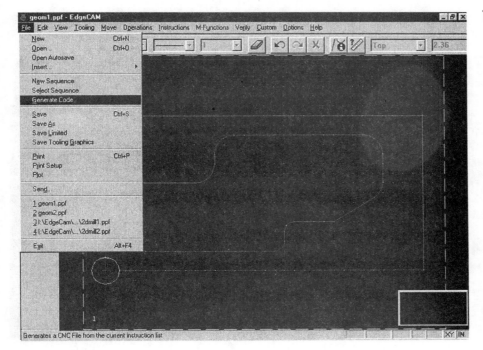

FIGURE 7.38

FIGURE 7.39

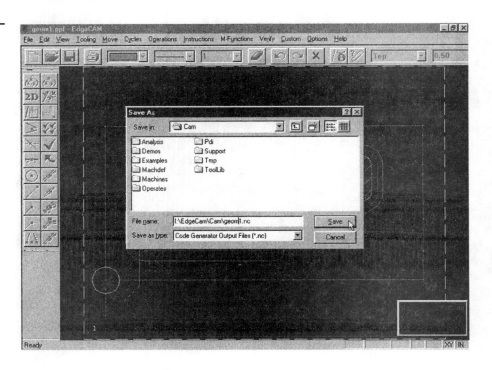

Enter a filename of geom1.nc then click Save (Fig. 7.39).

The code generator will prompt if you wish to display the NC code as it is generated as well as for programmer name and program number. The NC code will then be scrolled in a text window, and upon completion you will be prompted with some statistical information detailing that the CNC program is complete (Fig. 7.40).

The CNC program listing generated by this EdgeCAM example is included in the *EdgeCAMDemo* subfolder of the

FIGURE 7.40

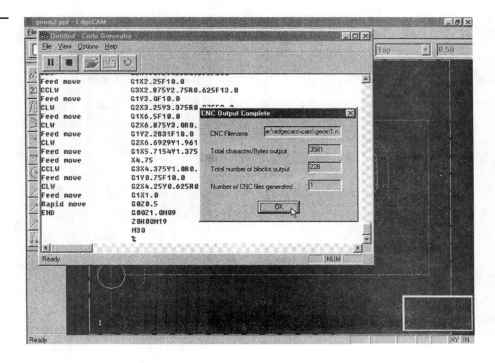

CNCWORK folder. You can run this code using the CNCez
Milling Simulator. The completed CNC program follows:

```
%
:1234
N10 G90 G70 G40 G17
N20 G00 X0.0 Y0.0
N30 Z1.0
N40 X6.0 Y0.0
N50 Z0.6
N60 T01 M06
N70 S1200 M03
N80 G00 X6.6408 Y2.124
N90 Z0.25 M08
N110 G01 X6.6408 Y2.124 Z-0.75 F5.0
N120 X6.6591 Y2.1423 Z-0.75 F15.0
N130 G03 X6.7125 Y2.2831 I-0.1591 J0.1408
N140 G01 X6.7125 Y2.3531 Z-0.75
N150 X6.2367 Y1.8773 Z-0.75
N160 X5.6704 Y1.5375
N170 X5.5787 Y1.5375
N180 X6.7125 Y2.6713
N190 X6.7125 Y2.9895
N200 X5.2605 Y1.5375
N210 X4.75 Y1.5375
N220 G03 X4.2125 Y1.0 I0.0 J-0.5375
N230 G01 X4.2125 Y0.7875 Z-0.75
N240 X4.1923 Y0.7875
N250 X4.2125 Y0.8077
N260 X4.2125 Y1.0
N270 X4.2327 Y1.1461
N280 X3.8741 Y0.7875
N290 X3.5559 Y0.7875
N300 X5.9809 Y3.2125
N310 X6.2991 Y3.2125
N320 X4.6039 Y1.5173
N330 X4.75 Y1.5375
N340 X4.9423 Y1.5375
N350 X6.5949 Y3.1901
N360 X6.5 Y3.2125
N370 X5.6627 Y3.2125
N380 X3.2377 Y0.7875
N390 X2.9195 Y0.7875
N400 X5.3445 Y3.2125
N410 X5.0263 Y3.2125
N420 X2.6013 Y0.7875
N430 X2.2831 Y0.7875
N440 X4.7081 Y3.2125
N450 X4.3899 Y3.2125
N460 X1.9649 Y0.7875
N470 X1.6467 Y0.7875
N480 X4.0717 Y3.2125
N490 X3.7535 Y3.2125
```

```
N500 X2.9684 Y2.4273
N510 G03 X3.0375 Y2.75 I-0.7184 J0.3227
N520 G01 X3.0375 Y2.8147 Z-0.75
N530 X3.4353 Y3.2125
N540 X3.25 Y3.2125
N550 G03 X3.0375 Y3.0 I0.0 J-0.2125
N560 G01 X3.0375 Y2.75
N570 G02 X2.5727 Y2.0316 I-0.7875 J0.0
N580 G01 X1.3285 Y0.7875 Z-0.75
N590 X1.0103 Y0.7875
N600 X2.1853 Y1.9625
N610 X1.8671 Y1.9625
N620 X0.8098 Y0.9052
N630 X0.7875 Y01.0
N640 X0.7875 Y1.2011
N650 X1.5489 Y1.9625
N660 X1.2307 Y1.9625
N670 X0.7875 Y1.5193
N680 X0.7875 Y1.75
N690 G02 X0.8409 Y1.8908 I0.2125 J0.0
N700 G01 X0.8592 Y1.9091 Z-0.75
N710 G00 X0.8592 Y1.9091
N720 X01.0 Y0.7875 Z0.5
N730 X01.0 Y0.7875 Z0.125
N740 G01 X01.0 Y0.7875 Z-0.75 F5.0
N750 G02 X0.7875 Y1.0 I0.0 J0.2125 F15.0
N760 G01 X0.7875 Y1.75 Z-0.75
N770 G02 X1.0 Y1.9625 I0.2125 J0.0
N780 G01 X2.25 Y1.9625 Z-0.75
N790 G03 X3.0375 Y2.75 I0.0 J0.7875
N800 G01 X3.0375 Y3.0 Z-0.75
N810 G02 X3.25 Y3.2125 I0.2125 J0.0
N820 G01 X6.5 Y3.2125 Z-0.75
N830 G02 X6.7125 Y3.0 I0.0 J-0.2125
N840 G01 X6.7125 Y2.2831 Z-0.75
N850 G02 X6.6093 Y2.1009 I-0.2125 J0.0
N860 G01 X5.6704 Y1.5375 Z-0.75
N870 X4.75 Y1.5375 Z-0.75
N880 G03 X4.2125 Y1.0 I0.0 J-0.5375
N890 G01 X4.2125 Y0.7875 Z-0.75
N900 X01.0 Y0.7875 Z-0.75
N910 G00 X01.0 Y0.7875 Z0.5
N920 M00
N930 X01.0 Y0.7875 Z0.0
N940 X6.0 Y0.0 Z0.0
N950 G00   Z1.0 M09
N960 M05
N970 T02 M06
N980 S1200 M03
N990 G00 X01.0 Y0.625
N1000 Z0.0 M08
N1010 M01
```

```
N1020 G01 X01.0 Y0.625 Z-0.75 F5.0
N1030 G02 X0.625 Y1.0 I0.0 J0.375 F10.0
N1040 G01 X0.625 Y1.75 Z-0.75
N1050 G02 X1.0 Y2.125 I0.375 J0.0
N1060 G01 X2.25 Y2.125 Z-0.75
N1070 G03 X2.875 Y2.75 I0.0 J0.625
N1080 G01 X2.875 Y3.0 Z-0.75
N1090 G02 X3.25 Y3.375 I0.375 J0.0
N1100 G01 X6.5 Y3.375 Z-0.75
N1110 G02 X6.875 Y3.0 I0.0 J-0.375
N1120 G01 X6.875 Y2.2831 Z-0.75
N1130 G02 X6.6929 Y1.9615 I-0.375 J0.0
N1140 G01 X5.7154 Y1.375 Z-0.75
N1150 X4.75 Y1.375 Z-0.75
N1160 G03 X4.375 Y1.0 I0.0 J-0.375
N1170 G01 X4.375 Y0.75 Z-0.75
N1180 G02 X4.25 Y0.625 I-0.125 J0.0
N1190 G01 X01.0 Y0.625 Z-0.75
N1200 G00 X01.0 Y0.625 Z0.5
N1210 G00  Z1.0 M09
N1220 G90 X0 Y0 M05
N1230 M30
```

STEP 7: Verify the CNC program

At this point you can verify the NC code dynamically by running the EdgeCAM Verify program.

From the View menu select Verify Machining.

The EdgeCAM Verify program will open with the machine part you generated in the EdgeCAM Manufacture mode (Fig. 7.41).

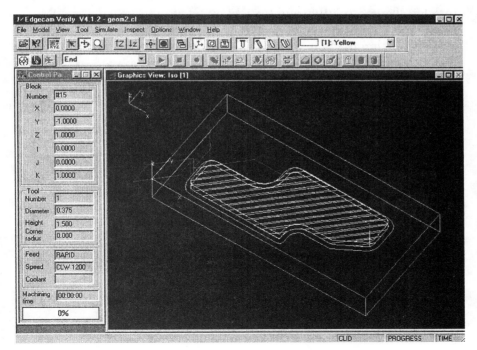

FIGURE 7.41

FIGURE 7.42
EdgeCAM Verify showing the rough pocketing operation.

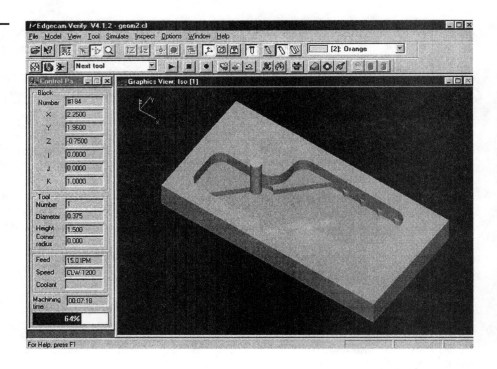

Select the Simulate item from the Simulate menu; then click the Start item of the Simulate menu to simulate the NC code dynamically (Fig. 7.42).

MASTERCAM

MasterCAM solutions are provided for two-dimensional, two and one-half–dimensional, and three-dimensional machining, including two-through five-axis milling and turning, two- and four-axis wire EDM, and sheet-metal punching and unfolding, plasma cutting, and lasers. All MasterCAM products have a double-precision, three-dimensional database, a powerful, integrated CAD system, and an easy-to-use intuitive user interface that helps users learn CAM quickly and easily.

Standard CAD features in MasterCAM include standard geometry and surface creation. The easy-to-use CAD system allows for the creation of the following entities in two- or three-dimensions: points, lines, arcs, fillets, splines, ellipses, rectangles, chamfers, and letters, as well as lofted, coons, ruled, revolved, swept, drafted, and trimmed surfaces. Also included are IGES, DXF, CADL, and ASCII bidirectional data converters. Other features include:

- Dimensioning in any plane or view
- Crosshatching
- Multiple viewports
- Dynamic rotation, panning, and zooming
- Plotting capabilities

CAM features in MasterCAM include the following:

- Graphic toolpath editing with full toolpath simulation
- Built-in tool libraries and materials files
- Canned cycle support
- Links to third-party applications
- Surface machining
- Drilling
- Pocketing
- Cycle time estimation

The following example shows the typical steps for creating geometry with MasterCAM and producing the final CNC code. For this example, the MasterCAM Version 5.55 milling module is used.

EXAMPLE OF CREATING GEOMETRY
IN MASTERCAM

After you enter the MasterCAM milling module, the main drawing screen appears (Fig. 7.43).

The MasterCAM screen consists of a graphics drawing area (center of screen), a menu area (left edge of screen), and a command area (bottom of screen).

When you move the mouse or pointing device, a cross hair or cursor moves around the graphics drawing area. By moving the cursor to the menu area on the screen's left edge, you can select a menu option. You

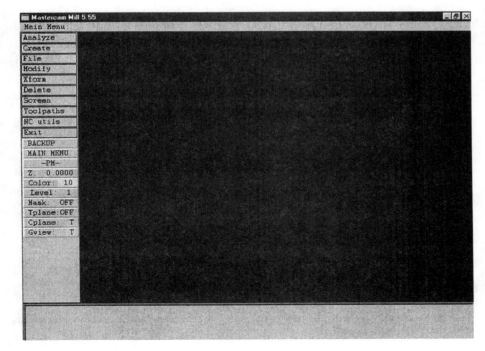

FIGURE 7.43
MasterCAM's graphic user interface.

FIGURE 7.44

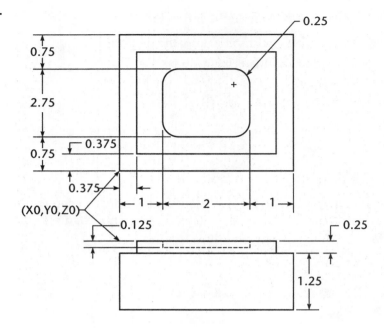

select an option by moving the cursor so that it highlights the option and then picking it with the pick button of your pointing device. You can also choose a menu option from the keyboard by typing in the capitalized letter shown for a particular menu option. For example, to choose File, press F on your keyboard.

Using MasterCAM and the part drawing in Fig. 7.44, generate the CNC part program that you can finally test on CNCez. (This is the same part drawing used in Exercise 7 in Chapter 8.)

STEP 1: Create a rectangle to show the stock material. First, create the rough stock. Remember that in order to start from a known position, the PRZ will be the lower left-hand top corner of the workpiece.

From the MAIN menu, select Create.

Select Rectangle.

Select 1 Point.

Select Values.

MasterCAM prompts you on the command line (bottom of screen) for the lower left-hand corner of the rectangle (Fig. 7.45).

Enter the following: X0Y0 [Enter].

When prompted for the width, enter 4 [Enter].

When prompted for the height, enter 3 [Enter].

You have now created a rectangle that will represent the stock of the workpiece to be machined (Fig. 7.46).

To see all the rectangle on the screen, use the Fit command: Press and hold Alt and press F1. The drawing will redraw

FIGURE 7.45

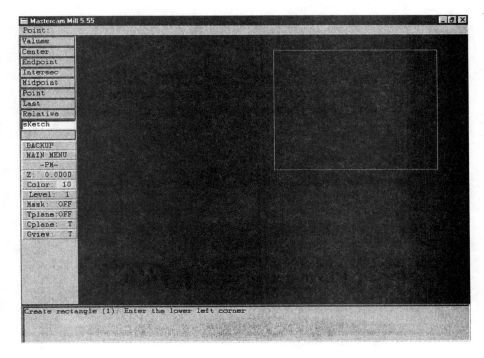

```
Mastercam Mill 5.55                                    _ 6 X




                    BACKUP
                    MAIN MENU
                     -PM-
                    Z:  0.0000
                    Color:  10
                    Level:  1
                    Mask:  OFF
                    Tplane:OFF
                    Cplane:  T
                    Gview:  T



Create rectangle (1): Enter the lower left corner
Enter coordinates:
X0Y0
```

automatically on the screen and fit itself to the entire screen area. The Fit command has just fit all objects on the screen; it has not changed the dimensions of the object. (See Fig. 7.47)

FIGURE 7.46

```
Mastercam Mill 5.55                                    _ 6 X
Point:
Values
Center
Endpoint
Intersec
Midpoint
Point
Last
Relative
sKetch

BACKUP
MAIN MENU
 -PM-
Z:  0.0000
Color:  10
Level:  1
Mask:  OFF
Tplane:OFF
Cplane:  T
Gview:  T




Create rectangle (1): Enter the lower left corner
```

FIGURE 7.47

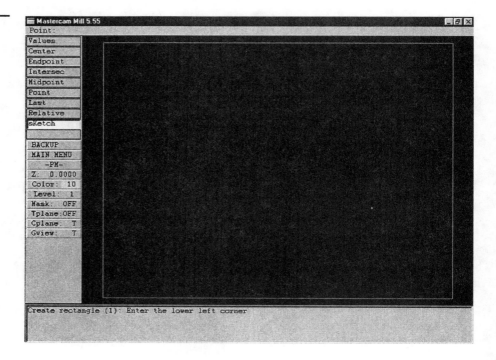

You can use the function keys to access MasterCAM macros that execute some of the more frequently used commands:

F1 = Zoom	Alt+F1 = Fit to screen
F2 = Unzoom	
F3 = Repaint	
F4 = Analyze	Alt+F4 = Cursor position
F5 = Delete	Alt+F5 = Delete window
F6 = File	
F7 = Modify	
F8 = Xform	
F9 = Display info	Alt+F9 = View
F10 = Help	

Another important key is Escape (Esc). Pressing it, you interrupt an operation and can back up one menu at a time.

STEP 2: Create a rectangle, which will be the basic shape of a lip around the outside of the part.

To return MasterCAM to the Main menu options, select Main Menu from the menu area.

You can return to the Main menu at any time by choosing Main Menu. To back up one menu at a time, choose the Back Up option.

From the Main menu, select Create.

Select Rectangle.

Select 1 Point.

Select Values.

MasterCAM prompts you on the command line for the lower left-hand corner of the rectangle.

Enter X.375Y.375 [Enter].

When prompted for the width, enter 3.25 [Enter].

When prompted for the height, enter 2.25 [Enter].

You have created a rectangle that will represent a raised ridge around the outside of the part. (Fig. 7.48).

Next, transform the rectangle that forms the lip to its final depth of cut value. In this example, the depth is –0.25 inch.

Return to the Main menu.

Select Xform.

Select Delete.

Select Translate.

Select Window.

Place a window around the inner rectangle (Fig. 7.49).

Select Done.

Select Rectangle.

Enter Z–0.25 [Enter].

At the dialog select Done (Fig. 7.50).

Press F3 to redraw the screen.

The lip is now a different color to show it has been translated to Z–0.25.

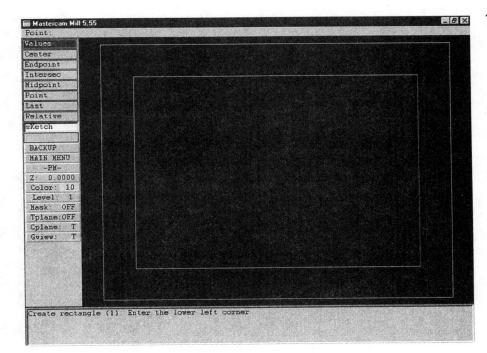

FIGURE 7.48

FIGURE 7.49

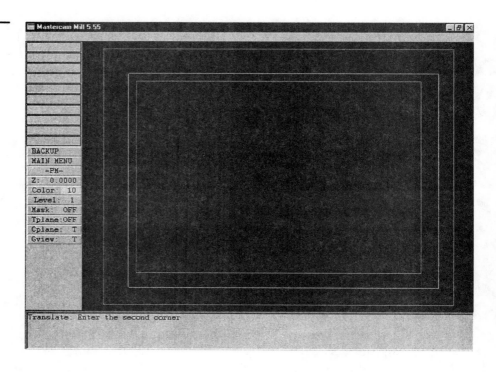

STEP 3:   Create a rectangle that will be the basic shape of an internal pocket.

From the Main menu, select Create.
Select Rectangle.
Select 1 Point.
Select Values.

FIGURE 7.50

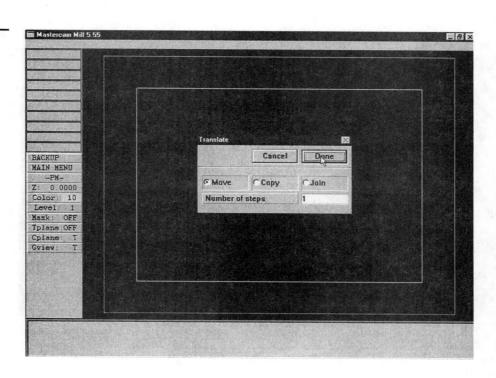

MasterCAM prompts you on the command line for the lower left-hand corner of the rectangle.

Enter X1Y0.75 [Enter].
Enter the width: 2 [Enter].
Enter the height: 1.5 [Enter].

You have created a rectangle that will represent the internal pocket (Fig. 7.51).

STEP 4:   Fillet the four corners of the internal pocket.

From the Main menu, select Modify.
Select Fillet.
Select Radius.
Enter the radius of the fillet: 0.25 [Enter] (Fig. 7.52).

In MasterCAM, the fillet radius must be larger than the radius of the cutter to be used.

You can now fillet the line of the first intersection, as shown in Fig. 7.53. After completing the first fillet, continue with the remaining intersections in the same manner.

Next, transform the pocket to its final depth of cut. In this example, the depth is −0.125 inch.

Return to the Main menu and select Xform.
Select Translate.
Select Window.

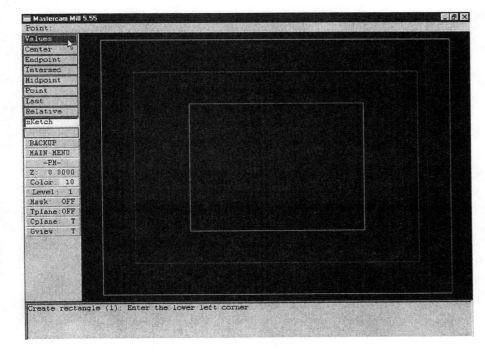

FIGURE 7.51

FIGURE 7.52

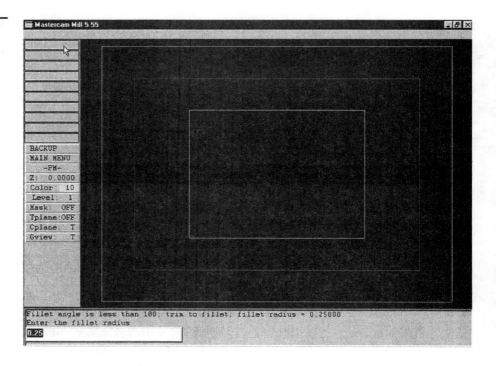

Create a window around the pocket (Fig. 7.54).

Select Done.

Select Rectangle.

Enter Z−0.125 [Enter].

At the Translate dialog click on Done.

Press F3 to redraw the screen.

FIGURE 7.53

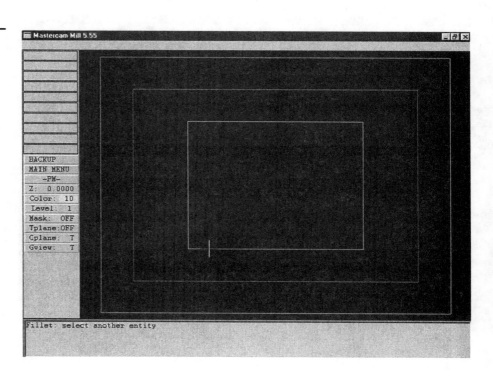

FIGURE 7.54

Mastercam Mill 5.55

BACKUP
MAIN MENU
—PM—
Z: 0.0000
Color: 10
Level: 1
Mask: OFF
Tplane:OFF
Cplane: T
Gview: T

Translate: Enter the second corner

STEP 5: Save the part geometry.

The geometry is now set at the proper depth for machining. You should save the geometry before creating the toolpaths, which is Step 6.

Return to the Main menu, then select File.
Select Save.
Enter Camexer1.ge3 [Enter] or click on Save (Fig. 7.55).

STEP 6: Next, create the toolpath. To machine this part:

1. Rough cut around and finish profiling the lip.
2. Cut out and finish the internal pocket.

You will use both the contour and the pocket toolpath commands. Use the contour toolpath for having the cutter follow the center of a line or follow tangent to that line (cutter compensation to the left or right of the line). Use the pocket toolpath to cut out a cavity with or without internal pockets.

From the Main menu, select Toolpaths.
Select Contour.

MasterCAM prompts you for an NCI file name. In MasterCAM, the toolpath file is referred to as an NCI file. The NCI file will be converted to CNC code for a particular machine tool, using the postprocessor.
Enter Camexer1 [Enter] or click on Save (Fig. 7.56).
Use the chain command to select the rectangle that forms the lip on the part.

FIGURE 7.55

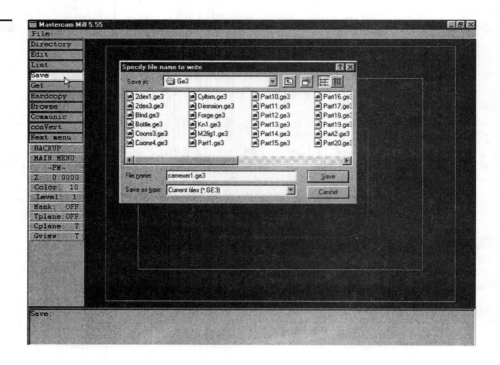

Select Chain.

Select the lower left-hand corner of the rectangle (Fig. 7.57).

To complete the chain, select Close.

The rectangle closes, and its color changes to show that the chain is complete.

To end the chained profile, select End Here (Fig. 7.58).

Select Done.

Next, set the machining parameters for this toolpath section.

Select Params.

Use the parameter dialog set without modification.

The parameter dialog have many different options; not all the parameters have to be changed to complete this toolpath. Definitions for each parameter are in the MasterCAM documentation.

The contour toolpath has two selections in the upper left of the parameter window. The first is specific to contouring; the second is generally the same for all NC toolpaths.

To move between the parameter screen options, click on the desired option and fill in the required fields, or click on the desired options. For the contour parameters set the following options as in Fig. 7.59:

Parameter file...	**CAMEXER1.PRM**
Contour Depth	**-0.25 Absolute**
Number of roughing cuts	**2**
Roughing cut spacing	**0.1**
Number of finish passes	**1**

FIGURE 7.56

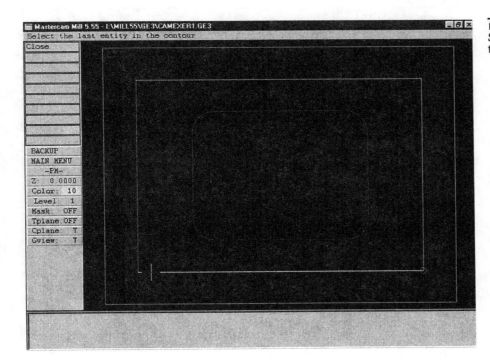

Finish pass spacing	0.05
Machine Finish Passes at	**All Depths**
Entry/exit line direction	**tangent**
Entry/exit line length	**0.0000**
Entry/exit arc: Radius	**0.0000**
Entry/exit arc: Angle	**90.000**
Infinite Look Ahead	**checked**

FIGURE 7.57
Selecting the contour entity for
the Chain command.

FIGURE 7.58
Finishing the Chain command.

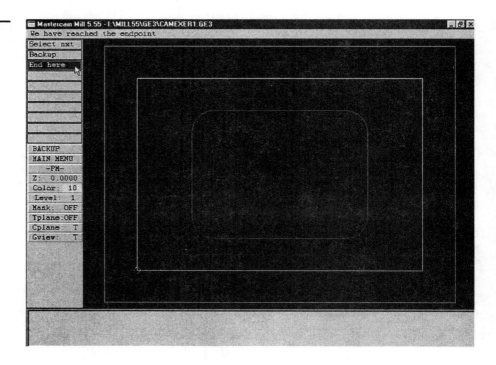

When these parameters are all entered, click on the NC option in the upper left of the dialog. The dialog will change to allow for configuring the NC options (Fig. 7.60).

FIGURE 7.59

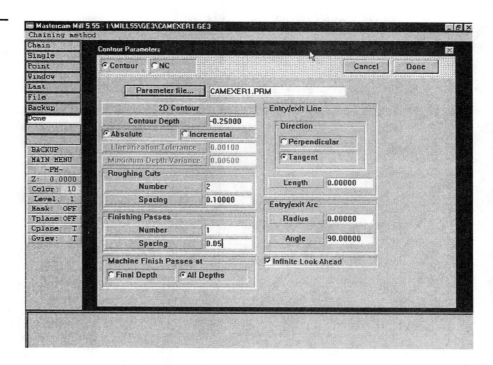

FIGURE 7.60

Contour Parameters

○ Contour   ● NC   [Cancel] [Done]

Tool Reference	02500FLT ▾

[Tool Library...]   [Material]

		TOOLS.MTL	NONE ▾
Tool number	2		
Diameter offset number	0		
Length offset number	0		
Cutter diameter	0.25000		
Corner radius	0.00000		
Stock to leave	0.00000		
Rapid Depth	0.12500		
Feedrate	8.00000		
Plunge rate	4.00000		
Spindle speed	2500		
Starting sequence number	5		
Increment	5		
Program number	1998		

Cutter Compensation

In computer	In control	Compensate to cutter
● Right	○ Right	● Center
○ Off	● Off	
○ Left	○ Left	○ Tip

Roll Cutter Around Corners
○ None   ● Sharp   ○ All

[Coolant]   [Flood ▾]

Depth cuts

2	rough cuts at	0.12500
0	finish cuts at	0.00000

Linear array

Nx	1	Dx	0.00000
Ny	1	Dy	0.00000

[Coordinates]   [Entry / Exit]   [Misc. Values]   [Tool Display]

Make the necessary changes so that the NC parameter screen matches the following parameters:

Tool Reference	**02500FLT**
Tool number	**2**
Diameter offset number	**0**
Length offset number	**0**
Cutter diameter	**0.25**
Corner radius	**0**
Stock to leave	**0**
Rapid Depth	**0.125**
Feedrate	**8**
Plunge rate	**4**
Spindle speed	**2500**
Starting sequence number	**5**
Increment	**5**
Program number	**1998**
Depth cuts **2** rough cuts at	**0.125**
0 finish cuts at	**0.00000**
Tool library:	**Tools.mtl**
Material:	**None**
Cutter compensation	
In computer	**Right**
In control	**Off**
Compensate to cutter	**Center**

FIGURE 7.61

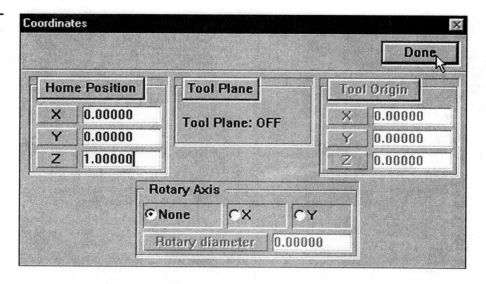

Roll Cutter Around Corners	**Sharp**
Coolant	**Flood**
Linear array: Nx, Ny	**1, 1**
Linear array:Dx, Dy	**0.0000, 0.0000**

Once all these parameters are entered, click the Coordinates button. Enter the following parameters and click Done when complete (Fig. 7.61).

Home position	
X	**0.0000**
Y	**0.0000**
Z	**1.0000**
Tool Plane:	**Off**
Tool Origin:	**Off**
Rotary Axis:	**None**

In the Contour Parameters dialog click the Tool Display button. A Tool Display dialog will appear. Enter the following parameters:

Display:	**Tool**
Toolpath	
Motion:	**Static**
Endpoints	
Run	
Delay	**0.0000**

When all the above parameters have been entered or selected, click Done (Fig. 7.62). Finally, in the Contour Parameters dialog click the Done button.

FIGURE 7.62

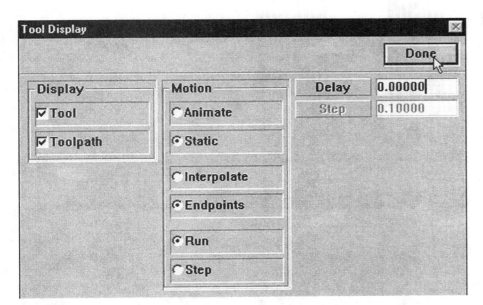

The rectangle should now have three light-blue cutter paths around the outside. If the cutter paths are to the inside, your chain direction has gone clockwise instead of counterclockwise. Return to the Main menu without saving the toolpath and try again.

To accept the toolpath segment, select Yes (Fig. 7.63).

The machining will now be continued by moving onto the pocket.

From the Toolpaths Menu, select Pocket (Fig. 7.64).

FIGURE 7.63

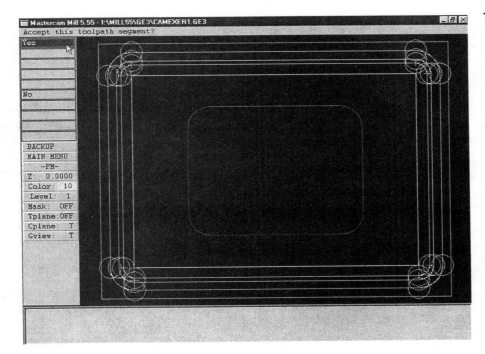

FIGURE 7.64

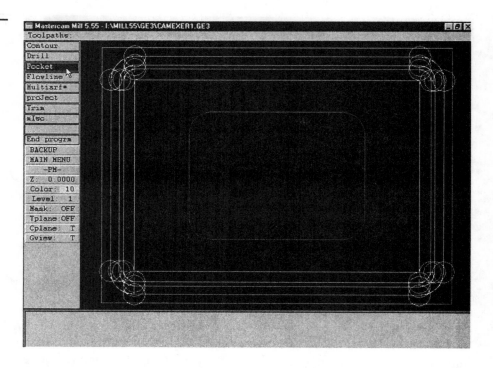

Use the Chain command to select the rectangle that forms the pocket.

Select Chain.

Select the lower left-hand corner of the pocket (Fig. 7.65).

In order to end the chain, select Close.

The rectangle closes, and its color changes to show that the chain is complete.

FIGURE 7.65

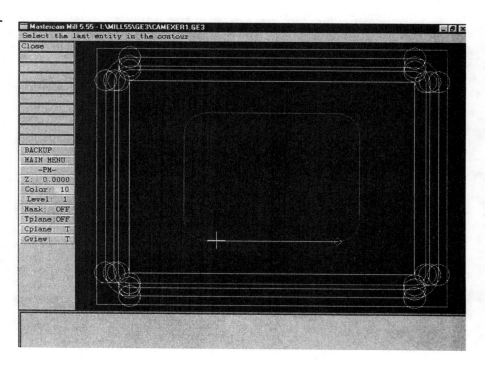

FIGURE 7.66

Pocketing parameters

○ Pocket   ○ NC   ○ Finish   Cancel   Done

☐ Batch mode   Parameter name   CAMEXER1.PRM

ROUGHING PARAMETERS
Roughing angle	0.00000	
Roughing cut spacing	0.10000	
Pocket depth	-0.25000	

Cutting method
○ One way - climb
○ One way - conventi.
● Zig zag
○ Spiral-out
○ Spiral-in

Helix/ramp entry [Off]
Tapered walls [off]
Break arcs [circles]
● Minimize tool burial
○ Minimize cutting time

FINISHING PARAMETERS

Finishing Passes
No. of passes	1
Finish pass spacing	0.00100

Entry/exit Line
Direction
● Tangent
○ Perpendicular

Machine finish passes after
○ all roughing operations
● each roughing operation

Machine finish passes at
● Final depth
○ All depths

Length   0.00000

Entry/exit Arc
Radius	0.00000
Angle	90.00000

☑ Optimize finish passes   ☑ Keep tool down
☐ Use additional finish parameters

To end the chained profile, select End Here.

Select Done.

Next, set the machining parameters for the pocket section (Fig. 7.66).

The Pocket parameters dialog has two parameter screens. The first is Pocket; the second NC. The NC screen, similar to all toolpath creations, is in the Contour Parameter steps described earlier.

Note that not all entries need to be changed to complete this toolpath. Definitions for each parameter are in the MasterCAM documentation.

Set and verify the following parameters in the Pocket dialog:

Roughing Parameters

Roughing angle	**0.000**
Roughing cut spacing	**0.100**
Pocket depth	**−0.125**
Cutting method	**Zig zag**

Finishing Parameters

Number of finish passes	**1**
Finishing pass spacing	**0.001**
Entry/exit line direction	**Tangent**
Entry/exit line length	**0.0000**
Entry/exit arc Radius	**0.0000**

FIGURE 7.67

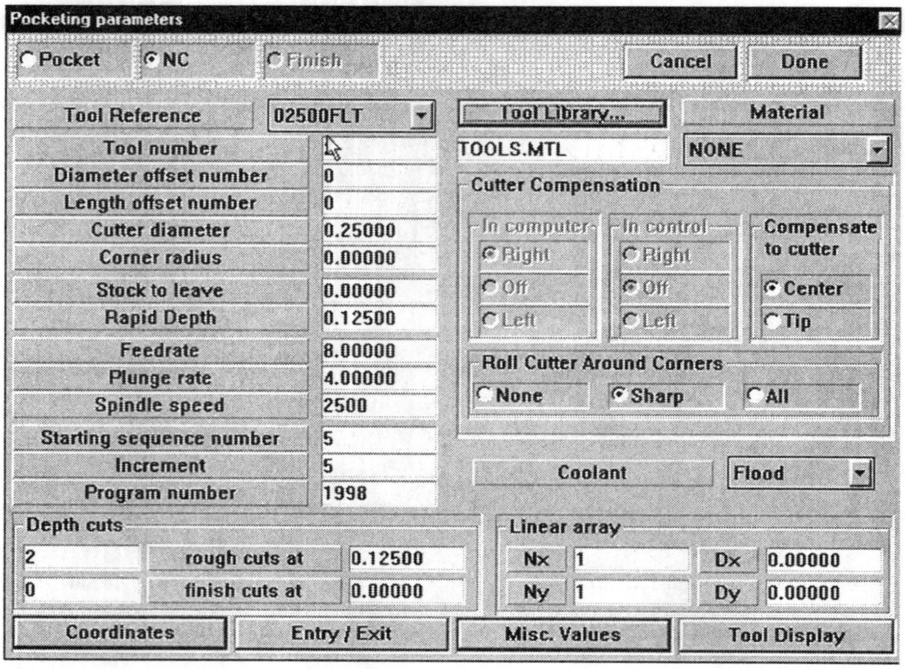

Entry/exit arc Radius Angle   **90.000**

Machine finish passes after   **each roughing operation**

Machine finish passes at   **Final depth**

After these parameters have been set, select the NC option in the top left of the dialog. The NC screen should display the NC parameters that were set in the Contour operation set previously. You can default these and click the Done button (Fig. 7.67).

The inside of the pocket should now have a white line following the inside of the pocket's shape.

> Select All. (Watch the toolpath move in the pocket.)
>
> Select Done.
>
> To accept the toolpath, select Yes.
>
> To end the toolpath generation, select End program.
>
> To run the postprocessor, click the Yes button (Fig. 7.68).

A text dialog, listing the generated CNC program (Fig. 7.69) will then be displayed.

The complete CNC program listing follows:

```
%
:1998
N001 G90 G80 G40
N002 M06 T02
N003 S2500 M03
N004 Z.125
N005 X.375 Y.100
N006 G01 Z-.125 F4.0
N007 X3.625 F8.0
```

FIGURE 7.68

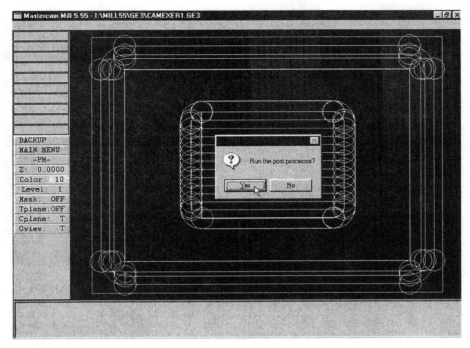

```
N008 G03 X3.900 Y.375 J0.275
N009 G01 Y2.625
N010 G03 X3.625 Y2.900 I-0.275
N011 G01 X.375
N012 G03 X.100 Y2.625 J-0.275
N013 G01 Y.375
N014 G03 X.375 Y.100 I0.275
N015 G01 Y.200 F4.0
N016 X3.625 F8.0
N017 G03 X3.800 Y.375 J0.175
N018 G01 Y2.625
```

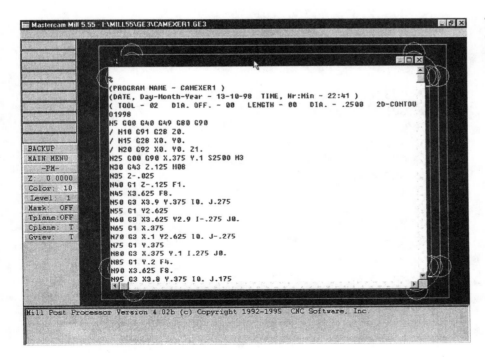

FIGURE 7.69
The completed CNC program listing.

```
N019 G03 X3.625 Y2.800 I-0.175
N020 G01 X.375
N021 G03 X.200 Y2.625 J-0.175
N022 G01 Y.375
N023 G03 X.375 Y.200 I0.175
N024 G01 Y.250 F4.0
N025 X3.625 F8.0
N026 G03 X3.750 Y.375 J0.125
N027 G01 Y2.625
N028 G03 X3.625 Y2.750 I-0.125
N029 G01 X.375
N030 G03 X.250 Y2.625 J-0.125
N031 G01 Y.375
N032 G03 X.375 Y.250 I0.125
N033 G00 Z.125
N034 Y.100
N035 G01 Z-.250 F4.0
N036 X3.625 F8.0
N037 G03 X3.900 Y.375 J0.275
N038 G01 Y2.625
N039 G03 X3.625 Y2.900 I-0.275
N040 G01 X.375
N041 G03 X.100 Y2.625 J-0.275
N042 G01 Y.375
N043 G03 X.375 Y.100 I0.275
N044 G01 Y.200 F4.0
N045 X3.625 F8.0
N046 G03 X3.800 Y.375 J0.175
N047 G01 Y2.625
N048 G03 X3.625 Y2.800 I-0.175
N049 G01 X.375
N050 G03 X.200 Y2.625 J-0.175
N051 G01 Y.375
N052 G03 X.375 Y.200 I0.175
N053 G01 Y.250 F4.0
N054 X3.625 F8.0
N055 G03 X3.750 Y.375 J0.125
N056 G01 Y2.625
N057 G03 X3.625 Y2.750 I-0.125
N058 G01 X.375
N059 G03 X.250 Y2.625 J-0.125
N060 G01 Y.375
N061 G03 X.375 Y.250 I0.125
N062 G00 Z.125
N063 X1.245 Y.876
N064 G01 Z-.125 F4.0
N065 X2.755 F8.0
N066 G03 X2.871 Y.972 I-0.005 J0.124
N067 G01 X1.129
N068 G02 X1.126 Y1.000 I0.121 J0.028
N069 G01 Y1.068
N070 X2.874
N071 Y1.164
N072 X1.126
```

```
N073 Y1.260
N074 X2.874
N075 Y1.356
N076 X1.126
N077 Y1.452
N078 X2.874
N079 Y1.548
N080 X1.126
N081 Y1.644
N082 X2.874
N083 Y1.740
N084 X1.126
N085 Y1.836
N086 X2.874
N087 Y1.932
N088 X1.126
N089 Y2.000
N090 G02 X1.129 Y2.028 I0.124
N091 G01 X2.871
N092 G03 X2.755 Y2.124 I-0.121 J-0.028
N093 G01 X1.245
N094 G02 X1.250 I0.005 J-0.124
N095 G01 X2.750
N096 G02 X2.874 Y2.000 J-0.124
N097 G01 Y1.000
N098 X2.875
N099 Y2.000
N100 G03 X2.750 Y2.125 I-0.125
N101 G01 X1.250
N102 G03 X1.125 Y2.000 J-0.125
N103 G01 Y1.000
N104 G03 X1.250 Y.875 I0.125
N105 G01 X2.750
N106 G03 X2.875 Y1.000 J0.125
N107 G00 Z.125
N108 G00 X0 Y0 Z1
N109 M05
N110 M30
```

THE CAD/CAM SYSTEM

CAD and CAM together create a direct link between product design and manufacturing. As discussed in this chapter, the CAD system is used to develop a geometric model of the part. This model then is used by the CAM system to generate part programs for CNC machine tools. The computer is the common element in both procedures. Both the CAD and CAM functions may be performed either by the same system or by separate systems located in different rooms or even different countries. The network of engineering, design, and manufacturing computers becomes the information highway that ties the CAD and CAM functions together.

Extending the connection between CAD and CAM to its logical limits within a company yields the concept of the computer-integrated enterprise

(CIE). In the CIE, all aspects of the enterprise are computer aided, from management and sales to product design and manufacturing. A manager can access geometric models, engineering analysis, and manufacturing quality control data for products. A design engineer can access market surveys, manufacturing process plans, and product life cycle data. A part programmer can access part geometric models, tool inventories, and current machine performance data. The technology that enables a CIE to work is the computer, which includes workstations or PCs available to all who need them, a network connecting all the workstations or PCs, and file servers to route data to where it is needed. In a networked environment, file and data standards are crucial to data exchange between workstations or PCs, and growth to newer technologies as they become available.

OVERALL BENEFITS OF CAD/CAM

Increased productivity is generally the justification for using a CAD/CAM system. Productivity increases with faster turnaround, better quality, and more accuracy.

CAD/CAM systems allow for rapid development and editing of designs and documentation. When a three-dimensional geometric model is produced in the design process, it then becomes a common element for engineering analysis, machining process planning (including CNC part programming), documentation (including engineering drawings), quality control, and so on. The coupling of CAD and CAM considerably shortens the time needed to bring a new product to market. More specifically, CAD/CAM will produce benefits in the following areas.

DESIGN

An efficient CAD system enables a designer to look at complex geometries and examine many different design alternatives. It also facilitates tedious operations such as maintaining standard part libraries, computing inertial (mass) properties, interfacing with analysis programs, and exchanging data with other software packages.

DRAFTING

Computer-aided drafting greatly facilitates the production, editing, storage, and plotting of complex engineering drawings. Edited drawings need only be replotted, not redrawn. Standard part libraries can be utilized. If the CAD system uses a three-dimensional geometric model as its primary data, the production of engineering drawings is a simple matter of selecting the desired views, watching the computer generate most of the drawing, selecting the desired dimensions, and annotating the appropriate features. Changes to the geometric model are automatically reflected in the engineering drawings.

MANUFACTURING

CAM provides the facility to generate CNC part programs directly from a three-dimensional geometric model or a two-dimensional engineering drawing. In addition, the CAM system tracks the machine capabilities, tool lists, materials properties, recommended feeds and speeds, and so on. The CAM system can also track many stored part programs and download them to a CNC machine tool as needed. Part programs can be verified offline using virtual machine tool simulations. Facilitating communication between the computer and the CNC machine greatly speeds up machining process planning.

MANAGEMENT CONTROL

Project management methods range from autocratic (one person makes all the decisions) to the democratic (design teams are formed for each product, and each team member has a say in all aspects of the product's design, production, and distribution). Whatever the management method used, the enhanced communication and data availability provided by a CAD/CAM system enables those who make decisions to have access to a wide range of data, from geometric models to part programs. Access to more information can lead to more-informed decisions. This is the essence of the CIE, whereby the computer network is used to collect, correlate, store, and distribute all the data used by a company, whether it be financial, marketing, engineering, manufacturing, quality control, or product liability data.

CONCURRENT ENGINEERING

Efficient CAD/CAM operations can facilitate concurrent engineering, which essentially is the running of many of the CAD and CAM functions in parallel. Concurrent engineering is designed to reduce product development time and thereby enhance competitiveness in a world economy. It relies heavily on communication, both personal and over computer networks, between all principal players in product development.

PRODUCT QUALITY

CAD/CAM helps improve overall product quality by providing for data storage and distribution, facilitating communication, decreasing product design time, increasing design process flexibility, facilitating change anywhere along the product design path, and allowing improved verification of part geometry and CNC part programs. The results of these benefits include the following:

- The design closely meets the product requirements, allowing the design goal to be met.
- Design and analysis time is reduced, hence shortening the time to bring the product to market.
- Production increases.
- Profits rise because of higher revenues, resulting from a better quality product and lower production costs.

LAB EXERCISES

1. What is CAD?

2. What are the two primary goals of CAD?

   (a)

   (b)

3. What is CAM?

4. What are the two main applications of CAM?

   (a)

   (b)

5. Does CAD/CAM establish a direct link between product design and manufacturing? Explain.

CHAPTER 8

Workbook Exercises

This chapter is intended to reinforce the concepts covered in Chapters 1–7 to help reaffirm the basic concepts of CNC. The exercises get progressively more difficult. They are also combined with a series of exercise questions to help understand the theoretical concepts. If you have problems with the exercises, you should go back through the chapters and review the material.

WORKBOOK EXERCISES

EXERCISE 1 A REVIEW

Answer the following questions.

1. What are the two axes of motion on a basic CNC lathe?

2. What G-code is used to set the PRZ for milling?

3. What are *miscellaneous functions?*

4. What are the three primary axes of motion on a basic CNC mill?

5. What coordinate system are you working in when you are programming from a fixed PRZ?

6. What is a *preparatory function?*

7. Are most G-codes considered to be modal commands?

8. What does *performing a dwell* mean?

9. How does a G01 and a G02 command differ?

10. In CNC milling, what is the purpose of cutter diameter compensation?

11. What does the *I* specify when a circular interpolation is programmed?

12. Which G-code is used to cancel any cutter compensation?

13. Give the letter address that corresponds to the following Letter Address descriptions:

Feedrate: _____

Spindle speed: _____

Block number: _____

Miscellaneous function: _____

Preparatory function: _____

X axis location: _____

Tool number: _____

EXERCISE 2 CALCULATING DIAMETRICAL COORDINATES FOR TURNING

Calculate the diametrical coordinates for the following turning part.

Coordinate Sheet

#	X	Z
1		
2		
3		
4		
5		
6		
7		
8		
9		
10		
11		
12		
13		
14		

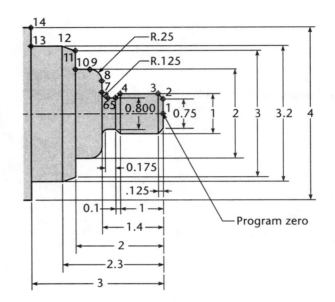

EXERCISE 3 CALCULATING COORDINATES FOR SIMPLE DRILLING AND SLOTTING

Calculate the coordinates for the following milling part.

Coordinate Sheet

#	X	Y	Z
1			
2			
3			
4			
5			
6			
7			
8			
9			
10			

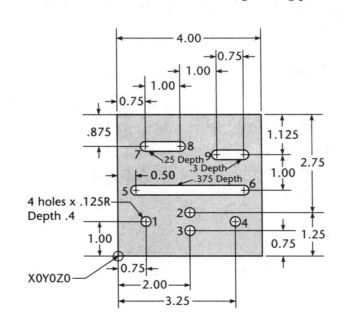

EXERCISE 4 LATHE PROFILE PROGRAMMING

Calculate the coordinates and complete the program for the finishing pass of the following lathe part. *Note:* This example is for demonstration purposes only. Actual machining is not advised.

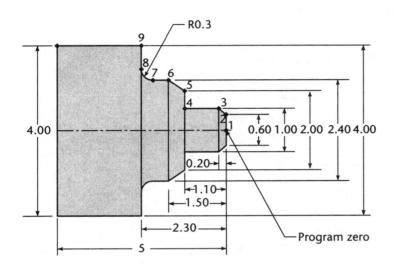

Coordinate Sheet

#	X	Z
1		
2		
3		
4		
5		
6		
7		
8		
9		

Workpiece Size: 4" Diameter by 5" Length

Tool: Tool #1, Right-hand Turning Tool

Tool Start Position: X4, Z3

```
%                              Program start flag
:1004                          Program number
N5  G__  G__  G__              Absolute, inches, comp. off
N10 T_____                   Tool change to Tool #1
N15 M0__                       Spindle on clockwise
N20 G0__  X0 Z0.1 M0__         Rapid to (X0, Z0.1), coolant 1 on
N25 G0__  Z__ F0.012           Feed to point #1 at 0.012 ipr
N30 X__                        Feed to point #2
N35 X__  Z__                   Feed to point #3
N40 Z__                        Feed to point #4
N45 X__                        Feed to point #5
N50 X__  Z__                   Feed to point #6
N55 Z__                        Feed to point #7
N60 G0__ X__ Z__ I0.3 K0       Circular feed to point #8
N65 G0__ X__                   Feed to point #9
N70 G0__ Z3 M0__               Rapid to Z3, coolant off
N75 T0100 M0__                 Tool comp. off, spindle off
N80 M0__                       Program end
```

EXERCISE 5 DRILLING BOLT HOLES

Calculate the drilling coordinates and complete the program for the following milling part. Use trigonometry to calculate the hole center points. Use the Workpiece/Save option to save the machined part for the next exercise.

Coordinate Sheet

#	X	Y	Z
1			
2			
3			
4			
5			
6			
7			
8			
9			
10			
11			
12			
13			
14			
15			

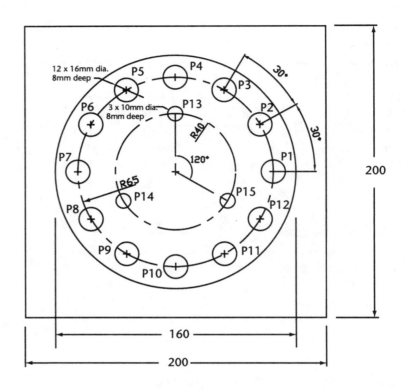

Workpiece Size: X200, Y200, Z40

Tool: Tool #1, 16mm Drill

              Tool #2, 8mm Drill

Tool Start Position: X0, Y0, Z40

%	Program start flag
:1005	Program number
N5 G__ G__ G40 G80	Absolute, mm, comp. and cycle off
N10 M____ T____	Tool change to Tool #1
N15 M0__ S_____	Spindle on clockwise at 1800 rpm
N20 G99 G__ X__ Y__ Z–8 R10 M0_	Start Drill cycle at point #1 (X_, Y_), coolant 2 on return to retract level of 2mm
N25 X__ Y__	Drill point #2
N30 X__ Y__	Drill point #3
N35 X__ Y__	Drill point #4
N40 X__ Y__	Drill point #5

N45 X__ Y__	Drill point #6
N50 X__ Y__	Drill point #7
N55 X__ Y__	Drill point #8
N60 X__ Y__	Drill point #9
N65 X__ Y__	Drill point #10
N70 X__ Y__	Drill point #11
N75 X__ Y__	Drill point #12
N80 G__ M0__	Cancel drill cycle, spindle off
M90 T0_ M06	Tool change select tool #2
M100 M__ S____	Spindle on, 2000 RPM
N105 G99 G81 X__ Y__ Z__ R__ F__	Spot drill point #13
N110 X__ Y__	Spot drill point #14
N115 X__ Y__	Spot drill point #15
N120 G80 M__	Cancel drill cycle, coolants off
N125 G__ Z1	Rapid to Z1 clearance
N130 X0 Y0 M__	Rapid to start, spindle stop
N135 M3__	Program end

EXERCISE 6 CIRCULAR ARC MOVES

This exercise emphasises the use of trigonometry to calculate the I and J values used in the G02 and G03 circular interpolation commands. Using the coordinates from Exercise 5, complete the CNC program for the following milling part, using G02 and G03 commands as indicated by A1–A4. Hint, use the workpiece machined from Exercise 5.

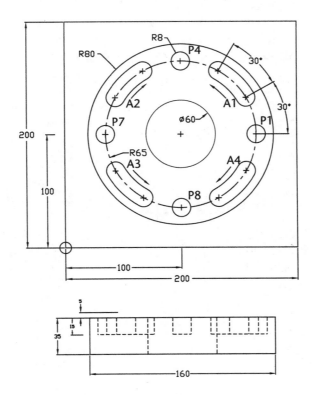

Workpiece Size: X200, Y200, Z40

Tool: Tool #1, 16mm Slot Drill

                     Tool #2, 30mm End Mill

Tool Start Position: X0, Y0, Z25

%	Program start flag
:1006	Program number
N5 G__ G__ G40 G80	Absolute, metric units, comp. and cycle off
N10 M__ T__	Tool change to Tool #1
N15 M__ S__	Spindle on at 1000 rpm
N20 G__ X__ Y__ M__	Rapid to startpoint of A1, coolant 2 on
N25 G__ Z___ F60	Feed into part at 60mm per min.
N30 G03 X__ Y__ I___ J___ F80	Perform 1st partial CCW arc
N35 G00 Z10	Rapid out to 10mm above part
N40 X__ Y__	Rapid to startpoint of A2
N45 G__ Z___ F60	Feed into part at 60mm per min.
N50 G02 X__ Y__ I__ J___ F80	Perform 2nd partial CW arc
N65 G00 Z10	Rapid out to 10mm above part
N70 X__ Y__	Rapid to startpoint of A3
N75 G01 Z___ F60	Feed into part at 60mm per min.
N80 G03 X__ Y__ I__ J___ F80	Perform 3rd partial CCW arc
N85 G00 Z10	Rapid out to 10mm above part
N90 X__ Y__	Rapid to startpoint of A4
N95 G01 Z___ F60	Feed into part at 60mm per min.
N100 G02 X__ Y__ I__ J___ F80	Perform 4th partial CW arc move
N110 G99 G81 X__ Y__ Z__ R__ F__	Drill a hole at P1
N115 X__ Y__	Drill a hole at P4
N120 X__ Y__	Drill a hole at P7
N125 X__ Y__	Drill a hole at P10
N130 G80 G00 Z__ M__	Rapid out to 25mm above part, coolant off
N135 T__ M__	Change to tool #2
N140 G00 X__ Y__ M__	Rapid to center of part, coolant on
N145 M__ S__	Spindle on, 800 rpm
N150 G01 Z-__ F60	Plunge into part 40mm
N155 X115	Feed to start of circle
N160 G03 X__ Y__ I__ J__	Cut a complete circle for center hole
N165 G00 Z__	Rapid to 25mm clearance
N170 X__ Y__ M__	Rapid to X0, Y0 home position Spindle off
N175 M__	Program end

EXERCISE 7 MILL PROFILING

Calculate the coordinates and complete the program for the following milling part.

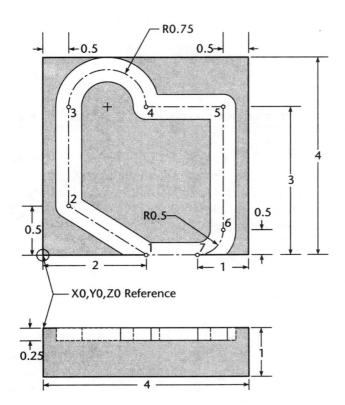

Coordinate Sheet

#	X	Y	Z
1			
2			
3			
4			
5			
6			
7			

Workpiece Size: X4,Y4,Z1

Tool: Tool #4, 1/2" Slot Drill

Tool Start Position: X0, Y0, Z1

%	Program start flag
:1007	Program number
N5 G__ G__ G40 G80	Absolute, inches, comp. and cycle off
N10 M____ T____	Tool change to Tool #4
N15 M0__ S_____	Spindle on clockwise at 2000 rpm
N20 G0__ X2 Y-0.375 M0__	Rapid to (X2, Y–0.375), coolant 2 on
N25 Z–0.25	Rapid down to Z–0.25
N30 G01 Y__ F15	Feed move to point #1 at 15 ipm
N35 X__ Y__	Feed move to point #2

N40 Y__	Feed move to point #3
N45 G02 X__ I0.75	Circular feed move to point #4
N50 G01 X__	Feed move to point #5
N55 Y__	Feed move to point #6
N60 G02 X__ Y__ I-0.5	Circular feed move to point #7
N65 G01 X__	Feed move to point #1
N70 G00 Z1	Rapid to Z1
N75 X0 M0__	Rapid to X0, coolant off
N80 M0__	Spindle off
N85 M3__	Program end

EXERCISE 8 MATERIAL REMOVAL FOR TURNING

Calculate the coordinates and complete the program for the following turning part. You must choose your tools from the Tool Library and properly calculate the speeds, feedrates, and depths of cut required. Try to utilize the G71 cycle to rough out the material. Be sure to use a Coordinate Sheet and plan your sequence of operations.

Workpiece Size: _____

Tool: _____

Tool Start Position: Z2, X3

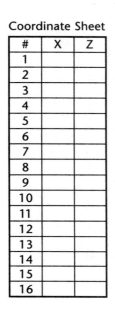

Coordinate Sheet

#	X	Z
1		
2		
3		
4		
5		
6		
7		
8		
9		
10		
11		
12		
13		
14		
15		
16		

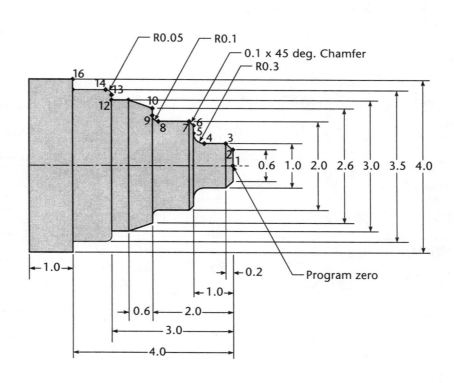

CNC PROGRAMMING SHEET		PART NAME:			PROGRAMMER:					
		MACHINE:			DATE:		PAGE:			
		SETUP INFORMATION:								

N SEQ	G Code	X Pos'n	Y Pos'n	Z Pos'n	I J K Pos'n		F Feed	R Radius or Retract	S Speed	T Tool	M Misc

N SEQ	G Code	X Pos'n	Y Pos'n	Z Pos'n	I J K Pos'n		F Feed	R Radius or Retract	S Speed	T Tool	M Misc

EXERCISE 9   MATERIAL REMOVAL FOR MILLING

Calculate the coordinates and create the program for the following milling part. Assume a 3/4-in. slot drill for the entire operation. You must calculate the feed rates and speeds based on the material you will be cutting. Be sure to use a Coordinate Sheet and plan your sequence of operations.

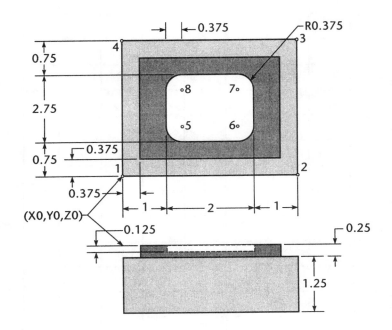

Coordinate Sheet

#	X	Y	Z
1			
2			
3			
4			
5			
6			
7			
8			

Workpiece Size: _____

Tool: _____

Tool Start Position: X0, Y0, Z1

CNC PROGRAMMING SHEET	PART NAME:		PROGRAMMER:		
	MACHINE:		DATE:	PAGE:	
	SETUP INFORMATION:				

N SEQ	G Code	X Pos'n	Y Pos'n	Z Pos'n	I J K Pos'n	F Feed	R Radius or Retract	S Speed	T Tool	M Misc

N SEQ	G Code	X Pos'n	Y Pos'n	Z Pos'n	I J K Pos'n		F Feed	R Radius or Retract	S Speed	T Tool	M Misc

ADVANCED EXERCISES

EXERCISE 10

The following exercise involves programming multiple turning operations, including drilling, facing, turning, and grooving. Calculate the coordinates and complete the program. You must choose your tools from the Tool Library and properly calculate the speeds, feedrates, and depths of cut required. Try to utilize the G71 cycle to rough out the material. Be sure to use a Coordinate Sheet and plan your sequence of operations.

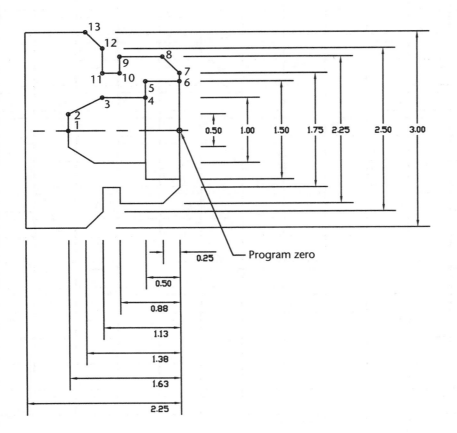

Coordinate Sheet

#	X	Z
1		
2		
3		
4		
5		
6		
7		
8		
9		
10		
11		
12		
13		
14		
15		
16		

Workpiece Size: _____

Tool: _____

Tool Start Position: Z2.25, X3

CNC PROGRAMMING SHEET		PART NAME:			PROGRAMMER:					
		MACHINE:			DATE:		PAGE:			
		SETUP INFORMATION:								
N SEQ	G Code	X Pos'n	Y Pos'n	Z Pos'n	I J K Pos'n		F Feed	R Radius or Retract	S Speed	T Tool	M Misc

N SEQ	G Code	X Pos'n	Y Pos'n	Z Pos'n	I J K Pos'n		F Feed	R Radius or Retract	S Speed	T Tool	M Misc

EXERCISE 11

The following milling part requires some advanced programming skills. Use radius compensation to make it easier to program the outer profile. Calculate the coordinates and complete the program. You must choose your tools from the Tool Library and properly calculate the speeds, feedrates, and depths of cut required. Be sure to use a Coordinate Sheet and plan your sequence of operations.

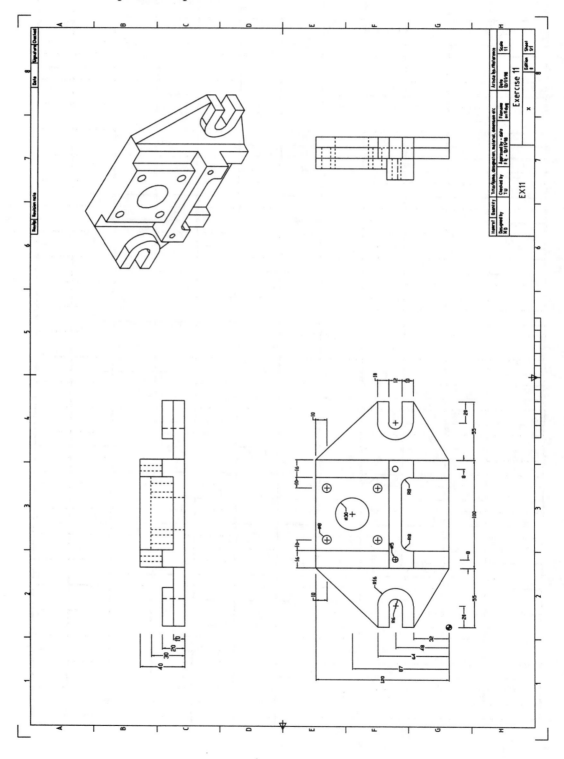

CNC PROGRAMMING SHEET		PART NAME:		PROGRAMMER:				
		MACHINE:		DATE:		PAGE:		
		SETUP INFORMATION:						

N SEQ	G Code	X Pos'n	Y Pos'n	Z Pos'n	I J K Pos'n		F Feed	R Radius or Retract	S Speed	T Tool	M Misc

N SEQ	G Code	X Pos'n	Y Pos'n	Z Pos'n	I J K Pos'n	F Feed	R Radius or Retract	S Speed	T Tool	M Misc

N SEQ	G Code	X Pos'n	Y Pos'n	Z Pos'n	I J K Pos'n		F Feed	R Radius or Retract	S Speed	T Tool	M Misc

ANSWERS TO LAB EXERCISES

Chapter 1: Lab Exercises

1. *What is CNC?*

   Computer numerical control, the process of manufacturing machined parts as controlled and allocated by a computerized controller.

2. *How did CNC come to be developed?*

   As a direct result of the war effort in World War II to ensure that both product quality and quantity requirements were met.

3. *What is DNC?*

   Direct numerical control, the process by which a computer directly controls a numerical control machine tool.

4. *List the steps in the CNC process.*

   1. Develop the part drawing.
   2. Decide which machine will produce the part.
   3. Choose the tooling required.
   4. Decide on the machining sequence.
   5. Do math calculations for the program coordinates.
   6. Calculate the speeds and feeds required for the tooling and part material.
   7. Write the NC program.
   8. Prepare set-up sheets and tool lists.
   9. Send the program to the machine.
   10. Verify the program.
   11. Run the program if no changes are required.

5. *Name some objectives of CNC.*

   1. Increase production
   2. Improve quality
   3. Improve accuracy
   4. Stabilize manufacturing costs
   5. Provide the ability to do more complex jobs

6. *What are some characteristics of CNC-produced parts?*

   1. Be similar in terms of raw material.

   2. Be of various sizes and shapes.

   3. Be a small to medium batch.

   4. Have parts whose sequences of steps for completing them are similar.

7. *Describe in your own words the CNC process.*

   NC programming is much the same as conventional machining. The machinist still has sole responsibility for the operation of the machine, although this control is no longer via manual turning of the axis handwheels, but rather via the controller and efficient NC planning and programming.

Chapter 2: In-Text Exercises

Milling Exercises

Exercise 1: Absolute Positioning

A. X–1.25, Y2	B. X–4, Y2	C. X–4, Y0.5
D. X–4.5,Y–2.5	E. X–2.5,Y–4.5	F. X3,Y–3.5
G. X4,Y–2	H. X4,Y2	

Exercise 2: Incremental Positioning

A. X–1.25, Y2	B. X–2.75, Y0	C. X0, Y–1.5
D. X–0.5, Y–3	E. X2, Y–2	F. X5.5, Y1
G. X1, Y1.5	H. X0, Y4	

Turning Exercises

Exercise 1: Using incremental coordinates, find the diametrical X and Z points on the profile.

A. X4, Z2	B. X0, Z–6	C. X–4, Z1
D. X0, Z–6	E. X2, Z–2	

Exercise 2: Using absolute coordinates, find the X and Z points on the profile.

A. X4, Z2	B. X4, Z–4	C. X2, Z–3
D. X2, Z–9	E. X3, Z–12	

Chapter 2: Lab Exercises

1. *What is the standard coordinate system called?*

   Cartesian coordinate system

2. *What are the three axes used on the CNC mill?*

X, Y, and Z

3. *What are the two axes used on the CNC lathe?*

X and Z

4. *What are the two types of coordinate systems? Explain the difference between them.*

The absolute and the incremental coordinate systems. The absolute system has a fixed origin, the incremental system does not (uses the last position as a reference).

5. *Does the X axis on a CNC milling machine run vertically or horizontally?*

Horizontally

6. *What are the three planes in the Cartesian coordinate system?*

XY, XZ, and YZ

7. *What is the PRZ?*

Program reference zero

8. *Where do you find the PRZ on the following?*

Milling workpiece: The lower left-hand and top corner surface of the part

Turning workpiece: The farthest Z-axis edge and center of the workpiece

Chapter 3: Lab Exercises

1. *What do the following letter addresses stand for?*

X: X coordinate

Y: Y coordinate

Z: Z coordinate

F: F assigns the feedrate

2. *What are the basic definitions of the following letter addresses?*

G: Preparatory function

M: Miscellaneous function

3. *What is a preparatory function?*

Functions for control of the machine that involve actual tool moves

4. *What are two reasons for using cutting fluids?*

1. Reduce heat

2. Reduce tool wear

3. Help in the removal of chips

5. *What factors affect how a cutting tool performs?*

   1. Tool material

   2. Shape of the tool point

   3. Form of the tool

   4. Cutting speeds and feedrates

6. *Describe the typical PRZ for milling.*

   The lower left-hand top corner of the work surface

Chapter 5: Lab Exercises

1. *What does the preparatory function G01 command do?*

   Linear interpolation command

   Moves the tool in a controlled linear fashion

2. *How are tool length offsets called up?*

   By using a G43 or G44

3. *Give an example of a rapid positioning in move.*

   G00 X1

4. *What does the address F stand for when a G01 command is programmed?*

   Linear feedrate

5. *What plane does a G18 specify?*

   XZ plane

6. *Which G-code and additional letter address are used to call up cutter compensation left?*

   G41 D_

7. *Write an example start line for a G81 drilling cycle.*

   N10 G81 Z–.25 R.125 F5

8. *Which M-code is used to specify program end, reset to start?*

   M30

Chapter 6: Lab Exercises

1. *What does the preparatory function G00 do?*

   Positioning in rapid

2. *How is tool nose radius compensation called up?*

   G41 or G42

3. *Give an example of a linear feed move.*

   G01 Z3

4. *What does the address U stand for when a G71 command is programmed?*

Amount of stock to be left for finishing in X

5. *What does a G76 command specify?*

Threading

6. *Which G-code and additional letter address are used to call up a dwell cycle?*

G04 P_

7. *Write an example start line for a G74 peck drilling cycle.*

N45 G74 Z–1 F.5 D.125 K.125

8. *Which M-code is used to specify spindle on clockwise?*

M03

Chapter 7: Lab Exercises

1. *What is CAD?*

Computer-aided design

2. *What are the two primary goals of CAD?*

(a) To increase productivity

(b) To create a database for manufacturing

3. *What is CAM?*

Computer-aided manufacturing

4. *What are the two main applications of CAM?*

(a) Computer directly controlling a manufacturing operation

(b) Computer used to support the manufacturing process

5. *How does CAD/CAM establish a direct link between product design and manufacture?*

The CAM system utilizes CAD data to develop the database for manufacturing. It also helps in the manufacturing process by generating CNC programs from this database.

Chapter 8: Workbook Exercises

Workbook Exercise 1

1. *What are the two axes of motion on a basic CNC lathe?*

X and Z

2. *What G-code sets the PRZ for milling?*

G92

3. *What are miscellaneous functions?*

Actions necessary for machining but not for tool movement

4. *What are the three axes of motion on a basic CNC mill?*

X, Y, and Z

5. *What coordinate system are you working in when you are programming from a fixed PRZ?*

Absolute

6. *What is a preparatory function?*

Functions for control of the machine that involve actual tool moves

7. *Are most G-codes considered to be modal commands?*

Yes, most are.

8. *What does performing a dwell mean?*

A delay in the program's execution

9. *How do a G01 command and a G02 command differ?*

G01 is linear; the G02 is an arc.

10. *In CNC milling, what is the purpose of cutter diameter compensation?*

To offset the tool center path left or right

11. *What does the I specify when a circular interpolation is programmed?*

Incremental distance from the startpoint of an arc to the centerpoint in X

12. *Which G-code is used to cancel any cutter compensation?*

G40

13. *Give the letter address that corresponds to the following:*

Feedrate: F

Spindle speed: S

Block number: N

Miscellaneous function: M

Preparatory function: G

X-axis location: X

Tool number: T

Workbook Exercise 2

The coordinates for the turned part:

1. X0, Z0
2. X0.75, Z0
3. X1, Z−0.125
4. X1, Z−1
5. X0.8, Z−1.1
6. X0.8, Z−1.275
7. X1.05, Z−1.4
8. X1.5, Z−1.4
9. X2, Z−1.65
10. X2, Z−2
11. X3, Z−2
12. X3.2, Z−2.3
13. X3.2, Z−3
14. X4, Z−3

Workbook Exercise 3

The coordinates for the milled part:

1. X0.75, Y1, Z−0.4
2. X2, Y1.25, Z−0.4
3. X2, Y0.75, Z−0.4
4. X3.25, Y1, Z−0.4
5. X5, Y1.875, Z−0.375
6. X3.5, Y1.875, Z−0.375
7. X0.75, Y3.125, Z−0.25
8. X1.75, Y3.125, Z−0.25
9. X2.75, Y2.875, Z−0.3
10. X3.5, Y2.875, Z−0.3

Workbook Exercise 4

The coordinates:

1. X0, Z0
2. X0.6, Z0
3. X1, Z−0.2

4. X1, Z−1.1

5. X2, Z−1.1

6. X2.4, Z−1.5

7. X2.4, Z−2

8. X3, Z−2.3

9. X4, Z−2.3

The completed program:

Workpiece Size:	4" Diameter by 5" Length
Tool:	Tool #1, Right-hand Turning Tool
Tool Start Position:	X4, Z3
%	Program start flag
:1004	Program number
N5 G90 G20 G40	Absolute, inches, comp. off
N10 T0101	Tool change to Tool #1
N15 M03	Spindle on clockwise
N20 G00 X0 Z0.1 M07	Rapid to (X0, Z0.1), coolant 1 on
N25 G00 Z0 F0.012	Feed to point #1 at 0.012 ipr
N30 X0.6	Feed to point #2
N35 X1 Z−0.2	Feed to point #3
N40 Z−1.1	Feed to point #4
N45 X2	Feed to point #5
N50 X2.4 Z−1.5	Feed to point #6
N55 Z−2	Feed to point #7
N60 G02 X3 Z−2.3 I0.3 K0	Circular feed to point #8
N65 G01 X4	Feed to point #9
N70 G00 Z3 M09	Rapid to Z3, coolant off
N75 T0100 M05	Tool comp. off, spindle off
N80 M02	Program end

Workbook Exercise 5

The coordinates:

1. X165, Y100, Z−8

2. X156.3, Y132.5, Z−8

3. X132.5, Y156.3, Z−8

4. X100, Y165, Z−8

5. X67.5, Y156.3, Z–8

6. X43.7, Y132.5, Z–8

7. X35, Y100, Z–8

8. X43.7, Y67.5, Z–8

9. X67.5, Y43.7, Z–8

10. X100, Y35, Z–8

11. X132.5, Y43.7, Z–8

12. X156.3, Y67.5, Z–8

13. X100, Y140, Z–10

14. X65.4, Y80, Z–10

15. X135.6, Y80, Z–10

Workpiece Size:	X200, Y200, Z40
Tool:	Tool #1, 16 mm Drill
	Tool #2, 10 mm Drill
Tool Start Position:	X0, Y0, Z25

The completed part program:

%	Program start flag
:1085	Program number
N5 G90 G21 G40 G80	Absolute, mm, comp., cycle off
N10 M06 T01	Tool change to Tool #1
N15 M03 S1800	Spindle on clockwise at 1800 rpm
N20 G99 G81 X165 Y100 Z–8 R2 F60	Start drill cycle at point #1, Retract level of 2 mm, coolant on
N25 X156.3 Y132.5	Drill point #2
N30 X132.5 Y156.3	Drill point #3
N35 X100 Y165	Drill point #4
N40 X67.5 Y156.3	Drill point #5
N45 X43.7 Y132.5	Drill point #6
N50 X35 Y100	Drill point #7
N55 X43.7 Y67.5	Drill point #8
N60 X67.5 Y43.7	Drill point #9
N65 X100 Y35	Drill point #10
N70 X132.5 Y43.7	Drill point #11
N75 X156.3 Y67.5	Drill point #12
N80 G80 M05	Cancel drill cycle, spindle off

N85 T02 M06	Tool change select Tool #2
N90 M03 S2000	Spindle on, 2000 rpm
N95 G99 G81 X100 Y140 Z–10 R2 F100	Spot drill point #13
N100 X65.4 Y80	Spot drill point #14
N105 X135.6 Y80	Spot drill point #15
N110 G80 M08	Cancel drill cycle, coolants off
N125 G00 Z1	Rapid to Z1 clearance
N130 X0 Y0 M05	Rapid to start, spindle stop
N135 M30	Program end

Workbook Exercise 6

Workpiece Size:	X200, Y200, Z40
Tools:	Tool #1, 16 mm Slot Drill
	Tool #2, 30 mm End Mill
Tool Start Position:	X0, Y0, Z25

The completed part program:

%

:1086

N5 G90 G21 G40 G80	Absolute, metric, comp., and cycle off
N10 M06 T01	Tool change to Tool #1
N15 M03 S1000	Spindle on at 1000 rpm
N20 G00 X156.2917 Y132.5 M08	Rapid to startpoint of A1 Coolant 2 on
N25 G01 Z-15 F60	Feed into part at 60 mm/min.
N30 G03 X132.5 Y156.3 I–56.2917 J–32.5 F80	Perform 1st partial arc
N35 G00 Z10	Rapid out to 10 mm above part
N40 X43.7083 Y132.5	Rapid to startpoint of A2
N45 G01 Z–15 F60	Feed into part at 60 mm/min.
N50 G02 X67.5 Y156.2917 I56.2917 J–32.5 F80	Perform 2nd partial arc
N65 G00 Z10	Rapid out to 10 mm above part
N70 X43.7083 Y67.5	Rapid to startpoint of A3
N75 G01 Z–15 F60	Feed into part at 60 mm/min.
N80 G03 X67.5 Y43.7083 I56.2917 J32.5 F80	Perform 3rd partial arc
N85 G00 Z10	Rapid out to 10 mm above part
N90 X156.2917 Y67.5	Rapid to startpoint of A4

N95 G01 Z–15 F60	Feed into part at 60 mm/min.
N100 G02 X132.5 Y43.7083 I–56.2917 J32.5 F80	Perform 4th partial arc move
N105 G00 Z25	Rapid to clearance of 25 mm
N110 G99 G81 X165 Y100 Z–15 R10 F50	Drill a hole at P1 15 mm deep retract to 10 mm above part
N115 X100 Y165	Drill a hole at P4
N120 X35 Y100	Drill a hole at P7
N125 X100 Y35	Drill a hole at P10
N130 G80 G00 Z25 M09	Rapid out to 25 mm above part coolant off
N135 T02 M06	Change to tool #2
N140 M03 S800	Spindle on at 800 rpm
N145 G00 X100 Y100 M08	Rapid to center of part coolant on
N150 G01 Z–40 F60	Plunge into part 40 mm
N155 X115	Feed to start of circle
N160 G03 X115 Y100 I–15 J0	Cut a complete circle for center hole
N165 G00 Z25	Rapid to 25 mm clearance
N170 X195 Y100	Profile the outside and clean up
N175 G01 Z–40	
N180 G03 I–95 J0	
N185 G01 X190 Y190	
N190 X195 Y195	
N195 X5	
N200 X15 Y185	
N205 X5	
N210 Y5	
N215 X15 Y15	
N220 Y5	
N225 X195	
N230 X190 Y15	
N235 X195	
N240 Y100	
N245 G00 Z25	
N250 X0 Y0 M09	Rapid to X0, Y0 home position Spindle off
N255 M30	End of program

Workbook Exercise 7

The coordinates:

1. X2.0, Y0, Z−.25

2. X.5, Y.5, Z−.25

3. X.5, Y3, Z−.25

4. X2, Y3, Z−.25

5. X3.5, Y3, Z−.25

6. X3.5, Y.5, Z−.25

7. X3, Y0, Z−.25

The completed part program:

Workpiece Size:	X4, Y4, Z1
Tool:	Tool #4, 1/2" Slot Drill
Tool Start Position:	X0, Y0, Z1
%	Program start flag
:1087	Program number
N5 G90 G70 G40 G80	Absolute, inches, comp., and cycle off
N10 M06 T04	Tool change to Tool #4
N15 M03 S2000	Spindle on clockwise at 2000 rpm
N20 G00 X2 Y−0.375 M08	Rapid to (X2, Y−0.375), coolant 2 on
N25 Z−0.25	Rapid down to Z−0.25
N30 G01 Y0 F15	Feed move to point #1 at 15 ipm
N35 X.5 Y.5	Feed move to point #2
N40 Y3.0	Feed move to point #3
N45 G02 X2 I0.75	Circular feed move to point #4
N50 G01 X3.5	Feed move to point #5
N55 Y.5	Feed move to point #6
N60 G02 X3 Y0 I−0.5	Circular feed move to point #7
N65 G01 X2	Feed move to point #1
N70 G00 Z1	Rapid to Z1
N75 X0 M09	Rapid to X0, coolant off
N80 M05	Spindle off
N85 M30	End of program

Workbook Exercise 8

The coordinates:

1. X0, Z0
2. X0.6, Z0
3. X1, Z–0.2
4. X1, Z–0.7
5. X1.6, Z–1
6. X1.8, Z–1
7. X2, Z–1.1
8. X2, Z–1.9
9. X2.2, Z–2
10. X2.6, Z–2
11. X3, Z–2.6
12. X3, Z–3
13. X3.4, Z–3
14. X3.5, Z–3.05
15. X3.5, Z–4
16. X4, Z–4

The completed part program:

Workpiece Size:	4" Diameter by 5" Length
Tools:	Tool #1, Right-hand Turning Tool
	Tool #2, Right-hand Finishing Tool
Tool Start Position:	X4, Z3
%	Program start flag
:1088	Program number
N5 G90 G20 G40	Absolute, inches, cutter comp. off
N10 T0101	Tool change to Tool #1
N15 M03	Spindle on clockwise
N20 G00 Z0.1 M07	Rapid to Z0.1, coolant #1 on
N25 G71 P30 Q105 U0.05 W0.05 D625 F0.012	Turning cycle
N30 G01 X0 Z0	Feed move to point #1
N35 X0.6	Feed move to point #2
N40 X1 Z-0.2	Feed move to point #3
N45 Z-0.7	Feed move to point #4
N50 G02 X1.6 Z–1 I0.3 K0	Circular feed move to point #5

N55 G01 X1.8	Feed move to point #6
N60 X2 Z-1.1	Feed move to point #7
N65 Z-1.9	Feed move to point #8
N70 G02 X2.2 Z-2 I0.1 K0	Feed move to point #9
N75 G01 X2.6	Feed move to point #10
N80 X3 Z-2.6	Feed move to point #11
N85 Z-3	Feed move to point #12
N90 X3.4	Feed move to point #13
N95 G03 X3.5 Z-3.05 I0 K-0.05	Circular feed move to point #14
N100 G01 Z-4	Feed move to point #15
N105 X4	Feed move to point #16
N110 G00 Z3 T0100	Rapid move to Z3
N115 T0202	Tool change to Tool #2
N120 G00 Z0.1	Rapid move to Z0.1
N125 G70 P30 Q105 F0.006	Finish pass on profile at 0.006 ipr
N130 G00 Z3 M09	Rapid to Z3 and coolant pump off
N135 T0200 M05	Tool cancel and spindle off
N140 M30	End of program

Workbook Exercise 9

The coordinates:

1. X0, Y0, Z-0.25
2. X4, Y0, Z-0.25
3. X4, Y3, Z-0.25
4. X0, Y3, Z-0.25
5. X1.375, Y1.125, Z-0.125
6. X2.625, Y1.125, Z-0.125
7. X2.625, Y1.875, Z-0.125
8. X1.375, Y1.875, Z-0.125

The completed part program:

Workpiece Size:	X4, Y3, Z1.5
Tool:	Tool #6, 3/4" Slot Drill
Tool Start Position:	X0, Y0, Z1
%	Program start flag
:1007	Program number
N5 G90 G20 G40 G80	Setup defaults

N10 M06 T6	Tool change to Tool #6
N15 M03 S1500	Spindle on clockwise at 1500 rpm
N20 G00 X−0.5 M08	Rapid move to X−0.5, coolant on
N25 Z−0.25	Rapid move down to Z−0.25
N30 G01 X4 F15	Feed move to point #2 at 15 ipm
N35 Y3	Feed move to point #3
N40 X0	Feed move to point #4
N45 Y0	Feed move to point #1
N50 G00 Z0.25	Rapid move up to Z0.25
N55 X1.375 Y1.125	Rapid move over to (X1.375, Y1.125)
N60 G01 Z−0.125 F5	Feed move down to point #5
N65 X2.625	Feed move to point #6
N70 Y1.875	Feed move to point #7
N75 X1.375	Feed move to point #8
N80 Y1.125	Feed move to point #5
N85 G00 Z1	Rapid move up to Z1
N90 X0 Y0 M09	Rapid move to (X0, Y0), coolant off
N95 M05	Spindle off
N100 M02	Program end

Workbook Exercise 10

Note: This is only one of many solutions to programming the part shown in Exercise 10.

Results will vary depending on tool selection, sequence, start positions, and roughing cycles selected.

Workpiece Size:	3" Diameter by 2.25" Length
Tools:	Tool #1, 1/8" Center Drill
	Tool #2, __" Drill
	Tool #3, __ Drill
	Tool #4, Internal Boring bar
	Tool #5, Right-hand Turn Tool
	Tool #5, Right-hand Finishing Tool
	Tool #7, 1/8" wide Groove Tool
Tool Start Position:	Z2, X3
Programming Mode:	Diametrical

%

:1811

N5 G20 G40 G98	(Inch programming, cancel comp, linear feedrate ipm)
N10 G97 T0101 M08	(Constant Spindle Speed, Tool change to #1, center drill, coolant on)
N15 S800 M03	(Spindle on, 800 rpm)
N20 G00 X0	(Rapid to Center line)
N25 Z.1	(Rapid to clearance)
N30 G01 Z−.150 F5.0	(Feed in .150 in. at 5 ipm)
N35 G00 Z.2	(Rapid clear)
N40 G28 Z1 T0100 M09	(Return to reference, cancel Tool #1, coolant off)
N45 T0202 M08	(Change to Tool #2, coolant on)
N50 G29 X0 Z.1	(Return from reference to clearance point)
N55 G01 Z−.5 F5.0	(Feed in .5 in. at 5 ipm)
N60 G00 Z.1	(Rapid clear)
N65 G28 Z1 T0200 M09	(Return to reference, cancel Tool #2, coolant off)
N70 T0303 M08	(Change to Tool #3, coolant on)
N75 G29 X0 Z.1	(Return from reference to clearance point)
N80 G74 Z−1.63 X0 D.150 K.150 F5.0	(Peck drill in 1.63 in)
N85 G28 Z1 M08	(Return to reference, coolant off)
N90 T0300 M05	(Cancel Tool #3, spindle off)
N95 G96 S600 M03	(Switch to Constant Surface Speed, spindle on)
N100 T0404 M08	(Change to Tool #4, coolant on)
N105 G99 G29 X0 Z.05	(Switch feedrate per rev., return from reference)
N110 G72 P115 Q135 U−.05 W0.05 D825 F0.015	(Rough face inside first)
N115 G01 Z.1 X1.5	(These blocks define inside profile for roughing, Point #6)
N120 Z−.5	(Point #5)
N125 X1.0	(Point #4)
N130 Z−1.13	(Point #3)
N135 X0.5 Z−1.63	(Point #2)

N140 G70 P115 Q135 F0.015 (Finish inside turning)

N145 G00 Z.1 (Rapid clear)

N150 G28 Z1 T0400 M09 (Return to reference, cancel Tool #4, coolant off)

N155 T0505 M08 (Change to Tool #5, coolant on)

N160 G00 Z.1 (Rapid to clearance)

N165 G71 P170 Q205 U.05 (Rough turn outside profile)
   W.05 D850 F0.015

N170 G01 X1.75 (These block define the outside profile, X of Point #7)

N175 Z0 (Z of Point #7)

N180 X2.25 Z−.25 (Point #8)

N185 Z−.88 (Point #9)

N190 Z−1.13 (Z of Point #12)

N195 X2.5 (Point #12)

N200 Z−1.38 X3.0

N205 G00 Z.1 (Rapid to clearance)

N210 G28 Z1 T0500 M09 (Return to reference, cancel Tool #5, coolant off)

N215 T0606 M08 (Change to Tool #6, coolant on)

N220 G29 Z.1 (Return from reference)

N225 G70 P170 Q205 F0.012 (Finish turn outside profile)

N230 G28 Z1 T0600 M09 (Return to reference, cancel Tool #6, coolant off)

N235 T0707 M08 (Select Tool #7, coolant on)

N240 G00 Z−1.005 (Rapid to startpoint in Z)

N245 X2.75 (Rapid to clearance point in X)

N250 G75 X1.75 Z−1.13 (Grooving cycle)
   I.125 K.125 D0 F0.015

N255 G28 Z1 M09 (Return to reference, coolant off)

N260 T0700 M05 (Cancel Tool #7, spindle off)

N265 M30 (Program end)

Workbook Exercise 11

Note: This is only one of many solutions to programming the part shown in Exercise 11.

Results will vary depending on tool selection, sequence, and start positions.

Workpiece Size:	X210, Y120, Z40
Tools:	Tool #1, 30 mm Face Mill
	Tool #2, 6 mm End Mill
	Tool #3, 16 mm End Mill
	Tool #4, 5 mm Drill
	Tool #5, 8 mm Drill
Tool Start Position:	X0, Y0, Z25

```
%
:1810
N5 G90 G80 G40 G21          (Safe start block)
N10 T01 M06                 (Tool change #1)
N15 M03 S1200                 (Set spindle on, 1200 rpm)
N20 G00 X55 Y−20              (Rapid out to startpoint)
N22 G01 G42 D10 Z−40 F8      (Profile outside with cutter comp.)
N25 Y0                        (Coordinates obtained from CAD drawing)
N30 Z−40
N35 X155
N40 X210 Y32
N45 Y64
N50 X155 Y120
N55 X55
N60 X0 Y64
N65 Y32
N70 X55 Y0
N80 Y−20
N85 G40 G00 Z25
N90 X−15 Y0
N95 G01 Z−30 F5                (Mill next face level)
N100 X40
N105 Y17
```

N110 X–15

N115 Y79

N120 X40

N125 Y135

N130 X170

N135 Y79

N140 X235

N145 Y17

N150 X170

N155 Y–15

N160 X124

N165 Y27   (Mill out front pocket)

N170 X86

N175 Y0

N180 X105

N185 G00 Z25

N190 X40 Y–5

N195 G01 Z–20 F5   (Mill the next face level)

N200 Y17

N205 X170

N210 Y0

N215 X55

N220 G00 Z25

N225 X170 Y–15

N230 G01 Z–20 F5

N235 Y42   (Mill the right side face level)

N240 X210

N245 Y60

N250 X170

N255 G00 Z25

N260 X40 Y17

N265 G01 Z–20 F5   (Mill the left side face level)

N270 Y42

N275 X0

N280 Y60

N285 X40

N290 Z25

N295 X55 Y135

N300 G01 Z–10 F5   (Mill out the top face level)

N305 Y69

N310 X70

N315 Y120

N320 X85

N325 Y69

N330 X100

N335 Y120

N340 X115

N345 Y69

N350 X130

N355 Y120

N360 X145

N365 Y69

N370 X150   (Clean up the cusps)

N375 X55

N380 G00 Z25

N385 X105 Y87

N390 G01 Z–40 F5   (Mill out the 30 mm hole)

N395 G00 Z25

N400 X210 Y120

N405 G01 Z–40 F5   (Clean up stubs for display)

N410 G00 Z25

N415 X0

N420 G01 Z–40 F5

N425 G00 Z25   (Rapid clear)

N430 X0 Y0 M05   (Rapid to home, spindle off)

N435 T02 M06   (Change Tool to #2, 6 mm slot mill)

N440 M03 S1200   (Spindle on, 1200 rpm)

N445 G00 X–10 Y45   (Rapid out to left side clear of part)

N450 G01 Z–40 F8   (Mill out left slot)

N455 X20

N460 G03 X20 Y51 I0 J3   (Use arc command to size slot end)

N465 G01 X−10

N470 G00 Z25

N475 X220

N480 G01 Z−40 F8

N485 X190   (Mill out right slot)

N490 G03 X190 Y45 I0 J−3   (Use arc command to size slot end)

N495 G01 X220

N500 G00 Z25   (Rapid clear)

N505 X0 Y0 M05   (Rapid to home, spindle off)

N510 T03 M06   (Tool change to Tool #3, 16 mm end mill)

N515 M03 S1200   (Spindle on 1200 rpm)

N520 G00 X55 Y130   (Position to left side)

N525 G01 Z−20 F5   (Mill out left side of top feature)

N530 Y62

N535 X63

N540 Y130

N545 X147

N550 Y62   (Mill out right side of top feature)

N555 X155

N560 Y130

N565 G00 Z25   (Rapid clear)

N570 X190 Y72   (Position for right side slot offset)

N575 G01 Z−30 F5   (Mill out outside offset of slot)

N580 G03 X190 Y24 I0 J−24   (Use arc for offset of slot end)

N585 G01 X163   (Feed back)

N590 Y72   (Clean up cusps)

N595 G00 Z25   (Rapid clear)

N600 X20   (Position for left side)

N605 G01 Z−30 F5   (Feed to Z level)

N610 G02 X20 Y24 I0 J−24   (Use arc to mill out offset of slot end)

N615 G01 X47   (Feed back)

N620 Y72   (Clean up cusps)

N625 G00 Z25   (Rapid clear)

N630 X79 Y24   (Position to inside pocket)

N635 G01 Z−30 F5   (Clean inside fillets)

N640 Y34	
N645 X131	
N650 G00 Z25	(Rapid clear)
N655 X0 Y0 M05	(Rapid home, spindle off)
N670 T04 M06	(Change to Tool #4, 5 mm drill)
N675 M03 S1200	(Spindle on 1200 rpm)
N680 G00 X63 Y48	(Position for topmost face holes)
N685 G81 Z–20 R10 F5	(Simple G81 drill cycle to Z–20)
N690 X147	(Second hole on topmost face)
N695 G80 G00 Z25 M05	(Cancel drill cycle, rapid clear, spindle off)
N700 T05 M06	(Change to Tool #5, 8 mm drill)
N705 M03 S1200	(Spindle on, 1200 rpm)
N710 G00 X129 Y64	(Rapid to first drill position)
N715 Z–5	(Rapid to clearance, caution here!)
N720 G99 G81 Z–20 R–8 F5	(Drill cycle with rapid to clearance plane!)
N725 Y110	(Drill out rest of the four holes)
N730 X81	
N735 Y64	
N740 G80 G00 Z25	(Cancel drill cycle, rapid to clearance)
N745 X0 Y0 M05	(Rapid home, spindle off)
N750 M30	(Program end)

CUTTING SPEEDS AND FEEDRATES

Formulas

Spindle rpm

$$\text{rpm} = \frac{4CS}{D}$$

CS   Material cutting speed in surface feet per minute/meters per minute (sfm/mpm)

D   Diameter of the part (turning) or diameter of the cutter (milling)

Feedrates (inch)

Lathe

Feed (in./min) = rpm $\times$ r

r   Feedrate in inches per revolution (ipr), (usually .001–.020 ipr)

Mill

Feed = rpm $\times$ T $\times$ N

T   Chip load per tooth

N   Number of teeth on cutter

Feedrates (metric)

Lathe

Feed (mm/min) = rpm $\times$ r

r   Feedrate in mm per revolution

Mill

Feed = rpm $\times$ T $\times$ N

T   Chip load per tooth

N   Number of teeth on cutter

MILLING MATERIAL TABLE (INCH VERSION)

(All cutting speeds in surface feet per minute, chip load per tooth)

Material	Cutting Speed	Chip Load
Aluminum and magnesium—HSS cutter	250	0.005
Brass and bronze, soft—HSS cutter	220	0.005
Copper—HSS cutter	150	0.005
Cast iron, soft—HSS cutter	75	0.005
Cast iron, hard—HSS cutter	50	0.003
Steel—HSS cutter	25	0.004
Stainless steel, hard—HSS cutter	35	0.003
Stainless steel, free machining—HSS cutter	70	0.003
Titanium—HSS cutter	35	0.003
Ferritic low alloys—HSS cutter	40	0.002
Austenitic alloys—HSS cutter	20	0.001
Nickel base alloys—HSS cutter	5	0.001
Cobalt base alloys—HSS cutter	5	0.001

MILLING MATERIAL TABLE (METRIC VERSION)

(All cutting speeds in surface meters per minute; chip load expressed in mm/flute)

Material	Cutting Speed	Chip Load
Aluminum and magnesium—HSS cutter	180	0.12
Brass and bronze, soft—HSS cutter	70	0.12
Brass and bronze, hard—HSS cutter	45	0.08
Copper—HSS cutter	45	0.12
Cast iron, soft—HSS cutter	23	0.12
Cast iron, hard—HSS cutter	14	0.07
Steel—HSS cutter	8	0.10
Stainless steel, hard—HSS cutter	11	0.08
Stainless steel, free machining—HSS cutter	21	0.08
Titanium—HSS cutter	11	0.08
Ferritic low alloys—HSS cutter	12	0.05
Austenitic alloys—HSS cutter	6	0.03
Nickel base alloys—HSS cutter	2	0.03
Cobalt base alloys—HSS cutter	2	0.03

LATHE MATERIAL FILE (INCH VERSION)

Column 1 is the spindle speed in in./min.

Column 2 is the feedrate in in./rev.

Column 3 is the feedrate for roughing only in in./rev.

Column 4 is the amount of each depth of cut in inches.

Column 5 is the cutting operation.

1	2	3	4	5
Aluminum				
300	0.007			Drilling
800	0.01	0.005	0.15	Roughing
1000	0.003	0.011		Finishing
600	0.003	0.011		Grooving
Steel				
120	0.005			Drilling
600	0.01	0.005	0.125	Roughing
800	0.003	0.01		Finishing
500	0.003	0.01		Grooving
Cast Iron				
120	0.003			Drilling
550	0.01	0.005	0.125	Roughing
410	0.003	0.011		Finishing
300	0.003	0.011		Grooving

LATHE MATERIAL FILE (METRIC VERSION)

Column 1 is the spindle speed in m/min.

Column 2 is the feedrate in mm/rev.

Column 3 is the feedrate for roughing only in mm/rev.

Column 4 is the amount of each depth of cut in mm.

Column 5 is the cutting operation.

1	2	3	4	5
Aluminum				
91	0.18			Drilling
244	0.25	0.12	3	Roughing
305	0.08	0.28		Finishing
183	0.08	0.28		Grooving
Steel				
37	0.13			Drilling
183	0.25	0.12	3	Roughing
244	0.08	0.25		Finishing
152	0.08	0.25		Grooving
Cast Iron				
37	0.08			Drilling
168	0.25	0.12	3	Roughing
125	0.08	0.28		Finishing
91	0.08	0.28		Grooving

APPENDIX C

Blank Programming Sheets

N SEQ	G Code	X Pos'n	Y Pos'n	Z Pos'n	I J K Pos'n	F Feed	R Radius or Retract	S Speed	T Tool	M Misc

PROGRAMMING SHEET

PART NAME:

MACHINE:

SETUP INFORMATION:

PROG BY:

DATE:

PAGE:

N SEQ	G Code	X Pos'n	Y Pos'n	Z Pos'n	I J K Pos'n	F Feed	R Radius or Retract	S Speed	T Tool	M Misc

PROGRAMMING SHEET

PART NAME:

MACHINE:

SETUP INFORMATION:

PROG BY:

DATE:

PAGE:

PROGRAMMING SHEET	PART NAME:					PROG BY:		
	MACHINE:					DATE:		PAGE:
	SETUP INFORMATION:							

N SEQ	G Code	X Pos'n	Y Pos'n	Z Pos'n	I J K Pos'n	F Feed	R Radius or Retract	S Speed	T Tool	M Misc

PROGRAMMING SHEET				PART NAME:							PROG BY:		
				MACHINE:							DATE:		PAGE:
				SETUP INFORMATION:									
N SEQ	G Code	X Pos'n	Y Pos'n	Z Pos'n	I J K Pos'n		F Feed	R Radius or Retract	S Speed	T Tool	M Misc		

PROGRAMMING SHEET

PART NAME:

MACHINE:

SETUP INFORMATION:

PROG BY:

DATE:

PAGE:

N SEQ	G Code	X Pos'n	Y Pos'n	Z Pos'n	I J K Pos'n		F Feed	R Radius or Retract	S Speed	T Tool	M Misc

PROGRAMMING SHEET

PROG BY:

PART NAME:

MACHINE:

DATE:

PAGE:

SETUP INFORMATION:

N SEQ	G Code	X Pos'n	Y Pos'n	Z Pos'n	I J K Pos'n	F Feed	R Radius or Retract	S Speed	T Tool	M Misc

PROGRAMMING SHEET

PART NAME: _____   PROG BY: _____

MACHINE: _____   DATE: _____

SETUP INFORMATION: _____   PAGE: _____

N SEQ	G Code	X Pos'n	Y Pos'n	Z Pos'n	I J K Pos'n	F Feed	R Radius or Retract	S Speed	T Tool	M Misc

PROGRAMMING SHEET

PART NAME:

MACHINE:

SETUP INFORMATION:

PROG BY:

DATE:

PAGE:

N SEQ	G Code	X Pos'n	Y Pos'n	Z Pos'n	I J K Pos'n		F Feed	R Radius or Retract	S Speed	T Tool	M Misc

PROGRAMMING SHEET

PART NAME:

MACHINE:

SETUP INFORMATION:

PROG BY:

DATE:

PAGE:

N SEQ	G Code	X Pos'n	Y Pos'n	Z Pos'n	I J K Pos'n	F Feed	R Radius or Retract	S Speed	T Tool	M Misc

PROGRAMMING SHEET

PART NAME:
MACHINE:
SETUP INFORMATION:

PROG BY:
DATE:
PAGE:

N SEQ	G Code	X Pos'n	Y Pos'n	Z Pos'n	I J K Pos'n	F Feed	R Radius or Retract	S Speed	T Tool	M Misc

TECHNICAL SUPPORT

GETTING TECHNICAL SUPPORT

If you have a problem or question about the software or a program you are writing, help is available from several sources. You may send a fax or write to the following:

TORCOMP Systems Ltd.
CNC Workshop Technical Support
7070 Pacific Circle
Mississauga, Ontario L5T 2A7
Canada
Fax: (905) 564-1377

You may send an e-mail to support@cncworkshop.com
You may also visit the CNC Workshop support page on the Internet at the following URL:

http://www.workshop.com   You will find a list of Frequently Asked Questions (FAQs) for the CNC Workshop version of *CNCez*.

When requesting assistance, make sure you provide us with as much information as possible, including the following:

1. **Address Information**
   Name
   Company
   Date
   Address
   Telephone number
   Fax number
   E-mail address

2. **Hardware and Software Information**
   Product name
   Computer brand and model, including CPU type
   Windows OS Version
   Network software and version
   Amount of hard disk space (used and free)
   Amount of memory (RAM)
   Type of graphics card
   Internet Explorer Version

3. **CNC Program**
   If the program is too long, attach it to an e-mail or mail it on a disk.

4. **Problem Description**
   The sequence of steps that produced the error
   Include Program and/or specific block numbers
   Error message displayed

GLOSSARY

Absolute system—A measuring system in which all points are given with respect to a common datum point. The alternative is the incremental system.

Accuracy—Measured by the difference between the actual position of the machine slide and the position demanded.

Address—A letter that represents the meaning of the element of information immediately following it.

ANSI—Abbreviation for American National Standards Institute. It sets drafting standards.

Arc clockwise—An arc generated by the tool motion in two axes in which the toolpath with respect to the workpiece is clockwise when the plane of motion is viewed from the positive direction of the perpendicular axis.

Arc counterclockwise—An arc generated by the tool motion in two axes, in which the toolpath with respect to the workpiece is counterclockwise, when the plane of motion is viewed from the positive direction of the perpendicular axis.

Array—A rectangular or circular pattern.

ASCII—Abbreviation for American Standard Code for Information Interchange. A standard set of 128 binary numbers that represent keyboard information such as letters, numerals, and punctuation.

Assembly—A drawing including more than one related part that is joined to or assembled with others.

Asynchronous port—An electrical connection port on a computer for one type of communication. Also called a serial or RS-232 communication (COM) port.

Attribute—Textual information associated with CAD geometry. Attributes can be assigned to drawing objects and extracted from the drawing database. Applications include creating bills of material.

Axis—A principal direction along which the movements of the tool or workpiece occur. There are usually three primary linear axes, mutually at right angles, designated X, Y, and Z.

Bezier curve—A polynomial curve passing near but not necessarily through a set of points. Represents an equation of an order one less than the number of points being considered.

Binary—The numerical base, base 2, by which computers operate. The electrical circuitry of a computer is designed to recognize only two states, high and low, which easily translate to logical and arithmetic values of 1 and 0. For example, the binary number 11101 represents the decimal number 29.

Bit—A binary digit (1 or 0). For example, the binary number 10110111 is eight bits long.

Bitmap—The digital representation of an image in which bits are referenced (mapped) to pixels. In color graphics, a different value is used for each red, green, and blue component of a pixel.

Block—A group of words that defines one complete set of instructions.

Block delete—Permits selected blocks of code to be ignored by the control system.

Bspline curve—A blended piecewise polynomial curve passing near a set of control points. The blending functions provide more local control as opposed to curves such as Bezier curves.

Buffer—An intermediate storage device (hardware or software) between data handling units.

Byte—A string of eight bits representing 256 binary values. A kilobyte (kbyte or KB) is 1024 bytes.

CAD—Abbreviation for computer-aided design, which uses graphics-oriented computer software for designing and drafting applications.

CAE—Abbreviation for computer-aided engineering, which uses graphics-oriented computer software for engineering and drafting applications involving mathematical analysis.

CAM—Abbreviation for computer-aided manufacturing, which is the use of computers to assist in various phases of manufacturing.

Cancel—A command that nullifies any canned cycles or sequence commands.

Cartesian coordinates—A means whereby the position of a point can be defined with reference to a set of axes at right angles to each other.

Chamfer—A beveled edge or corner between two intersecting lines or surfaces.

CIE—Abbreviation for computer-integrated engineering or computer-integrated enterprise.

CIM—Abbreviation for computer-integrated manufacturing, which involves a common computer database from which information for various manufacturing processes is stored and retrieved. This information usually includes drawings, numerical control data, and bills of materials.

Circular interpolation—Enables the programmer to move a tool up to 360✗ in an arc by using only one block of information. The circular path may be generated in any two planes.

CNC—Abbreviation for computer numerical control (see next page).

Code—A system that describes the assembling of characters for representing information.

Command—A signal or group of signals that initiates one step in the execution of a program.

Computer numerical control—(CNC) A numerical control system utilizing a computer as a controller.

Control key—A key on the keyboard used in conjunction with other keys to perform special functions.

Coons patch—A bicubic surface patch interpolated between four adjoining general space curves.

CPU—Abbreviation for central processing unit. The CPU controls, sequences, and synchronizes the activities of all the computer components and performs the various arithmetic and logic operations on the data.

Cross-hairs—A cursor usually comprises two perpendicular lines on the display screen and used to select coordinate locations.

CRT—Abbreviation for cathode ray tube. Denotes the video display tube used with computers.

Cursor—A pointer on a display screen that can be moved around and used to place textual or graphical information.

Cutter diameter compensation—Provides a means of using a cutter of a different diameter than originally intended in a program. The programmer may use either an oversized or undersized cutter and still maintain the programmed geometry.

Cutter offset—The distance from the part surface to the axial center of a cutter.

Cycle—A sequence of operations that is repeated regularly. The time it takes for one such sequence to occur.

Data—Information used as a basis for reasoning or calculation.

Database—Related information organized and stored so that it can be retrieved easily and, typically, used by multiple applications. A non-computer example of a database is a telephone directory.

Default—A parameter or variable that remains in effect until changed. It is what a computer program recognizes in the absence of specific user instructions.

Digitize—Act of entering graphic location points into a computer with a tablet, puck, or stylus.

Digitizing tablet—A graphics input device that generates coordinate data. It is used in conjunction with a puck or a stylus.

Directory—A portion of the storage space on a disk drive that can contain files. It is analogous to a file drawer in a filing cabinet.

Diskette—See **Floppy disk.**

Display screen—A video display tube or CRT used to transmit graphic information.

DOS—Abbreviation for disk operating system. Software that controls the operation of disk drives, memory usage, and I/O in a computer.

Drawing file—A collection of graphic data stored as a set (file) in a computer.

Dwell—A delay in a program's execution. A dwell halts all axis movement for a specified time.

DXF—Abbreviation for drawing interchange file. A file format used to produce an ASCII description of an AutoCAD drawing file.

Edit—To modify or prepare.

End point—The exact location where a line or curve terminates.

Enter—A keyboard key that when pressed signals the computer to execute a command or terminate a line of text. Also sometimes called Return.

Ethernet—The most widely used local area network technology, originally developed by Xerox. Currently specified by the IEEE802.3 standard, also known as 10Base-T.

FEA—Abbreviation for finite element analysis. Numeric technique of approximately determining field variables such as displacements and stresses in a domain. This is done by breaking the domain into a finite number of "pieces," also called "elements," and solving for the unknowns in those elements.

Feed—The programmed or manually established rate for movement of the cutting tool into the workpiece.

Feedrate (code word)—A code containing the letter F followed by digits. It determines the machine slide's rate of feed.

FEM—Abbreviation for finite element modeling. The process of breaking down a geometric model into a mesh, called the finite element mesh model, that is used for finite element analysis.

File—A collection of data accessible to a computer either on a disk drive or main memory that represents textual and/or graphic information.

Fillet—A curved surface of constant radius that blends two otherwise intersecting surfaces. A two-dimensional representation of this surface that involves two lines or curves and an arc.

Floppy disk—A circular plastic disk coated with magnetic material mounted in a square cardboard or plastic holder. Used by a computer to store information for later use. Can be inserted or removed from a floppy disk drive at will. Also called a diskette.

G-code—A word addressed by the letter G and followed by a numeric code that defines preparatory functions or cycle types in a numerical control system.

Generation, NC—Typically refers to the automatic generation of NC instructions from a CAD model of part geometry.

Grid—An area on the graphics display covered with regularly spaced dots, used as a drawing aid.

Hard disk—A rigid metal disk covered with magnetic material. Mounted permanently in a hard disk drive, it spins at a high velocity, is capable of storing large amounts of data, and works faster than a floppy disk drive does.

Hardcopy—A paper printout of information stored in a computer.

Hatching—A regular pattern of line segments that cover an area bounded by lines and/or curves.

Hexadecimal—A numbering system that contains 16 numbers as base units, also referred to as base-16 numbers. Hexadecimal numbers use numeral digits 0 through 9 and the letters from A through F. Hexidecimal is also used in modern computers to represent 2 bytes, or 16-bit, binary data.

Icon—A graphic symbol typically used to convey a message or represent a command on the display screen.

IGES—Abbreviation for International Graphics Exchange Standard. A standard of exchanging graphical data between CAD systems.

Incremental system—A control system whereby each coordinate or positioned dimension—both input and feedback—is taken from the last position rather than from a common datum point, as in the absolute system.

Input—External information entered into the control system.

Interface—A condition of spatial intersection between two parts in an assembly. Also, the region of intersection of the two parts.

Internet—A worldwide system of computer networks. Originally conceived by the U.S. government's Advanced Research Projects Agency (ARPA) in 1969, was called ARPAnet. Its intent was to construct a computer network that in the event of nuclear war would continue to function if parts of it were destroyed.

IPM—Abbreviation for inches per minute.

IPR—Abbreviation for inches per revolution.

ISO—Abbreviation for International Standardization Organization, an organization charged with establishing and promoting international standards.

Isometric—A view or drawing of an object in which the projections of the X, Y, and Z axes are spaced 120° apart and the projection of the Z axis is vertical.

Jog—A control function that manually operates an axis of the machine.

LAN—Abbreviation for local area network. One of several systems used to link computers in order to share data, programs, and peripherals.

Letter address—The manner by which information is directed to the system. All information must be preceded by its proper letter address—for example, X, Y, Z, or M.

Linear interpolation—The movement of the tool in a linear (straight) path.

Macro—A single command made up of a string of commands.

Mainframe computer—Arguably, a larger and faster computer than a minicomputer.

Manual data input—A method that enables an operator to insert data into the control.

Manual part programming—Programming method whereby the machining instructions are prepared by the operator on a document called the part program manuscript.

MCU—Abbreviation for machine control unit. Consists of the electronics and hardware that read and interpret the programmed instructions and convert them into the mechanical actions of the machine tool.

Memory—An essential component of a computer. The place in which programs and data are stored. Memory includes both ROM (read-only memory) and RAM (random-access memory).

Menu—A list of commands available for selection. Can be available on a digitizing tablet or on the display screen.

Microcomputer—A computer principally designed for use by a single person.

Microprocessor—An integrated circuit chip (or set of chips) that acts as the CPU of a computer—for example, the Motorola 68020 and the Intel 80386.

Minicomputer—A computer that is generally configured for simultaneous use by a small number of people. It generally has more power and peripherals than does a microcomputer.

Mirror—To create the reverse image of selected graphic items.

Modal—Information that once input into the system remains in effect until it is changed.

Mode—A software setting or operational state.

Model—A two- or three-dimensional representation of an object.

Modem—Stands for modulator/demodulator. The device that allows a computer to send and receive data over telephone lines.

Mouse—A hand-operated, relative-motion device that resembls a digitizer puck and is used to position the cursor on a computer display screen.

MPM—Abbreviation for millimeters per minute.

MPR—Abbreviation for millimeters per revolution.

Multitasking—The ability of an operating system to manage concurrent tasks on a computer.

Multiuser—The ability of an operating system to allow multiple users on different terminals to share computer resources such as the CPU, storage, and memory.

NC (numerical control)—The technique of controlling a machine or process by using numbers, letters, and symbols.

Network—An electronic linking of computers for communication.

Numerical control system (NC)—A system in which a program of instructions is read by a machine control unit and decoded to cause a machine tool movement or to control a process.

NURBS—Stands for Non-Uniform Rational Bsplines. A widely used parametric model for three-dimensional curves and surfaces.

Offset—A displacement in the axial direction of the tool that is the difference between the actual tool length and the programmed tool length.

Operating system—Also, disk operating system. Software that manages computer resources and allows a user access and control.

Origin—The intersection point of the axes in a coordinate system. For example, the origin of a Cartesian coordinate system is where the X, Y, and Z axes meet, at (0, 0, 0).

Orthogonal—Two geometric entities whose slopes or tangents are perpendicular at their intersection.

Orthographic projection—Also called the parallel projection, the two-dimensional representation of a three-dimensional object but without perspective. In drafting, it is typically the front, top, and right-side views of an object.

Part program—A specific and complete set of instructions for the manufacture of a part on an NC machine.

Part programmer—A person who prepares the planned sequence of events for the operation of a numerically controlled machine tool.

Peripheral—An accessory device to a computer such as a plotter, printer, or tape drive.

Pixel—Stands for "picture element" Pixels are the tiny dots that make up what is displayed on a CRT. Also called pels.

Plotter—A computer-controlled device that produces text and images on paper or acetate by electrostatic, thermal, or mechanical means (with a pen).

Point-to-point control system—A system in which the tool is moved to a predefined location. Only positioning is performed and there is no cutting performed during the positioning move. Also called a positioning system.

Preparatory function—An NC command that changes the mode of operation of the control. (Generally noted at the beginning of a block by the letter G and two digits.)

Program—A sequence of steps that is executed in order to perform a given function.

Prompt—A message from the computer software requesting a response from the user.

Puck—A hand-operated device with one or more buttons, resembling a mouse, that operates in conjunction with a digitizing tablet. Also called a transducer.

Quadrant—Any of the four parts into which a plane is divided by rectangular coordinate axes in that plane.

RAM—Abbreviation for random-access memory. The main memory of a computer. Programs and data can be read from and written to RAM.

Rapid—Positioning of the cutter near the workpiece at a high rate of travel speed before the cut is started.

Register—An internal memory storage location for the recording of information.

Relative coordinates—Coordinates specified by differences in distances and/or angles measured from a previous set of coordinates rather than from the origin.

Reset—To return a register to zero or to a specified initial condition.

Right-hand rule—A method of determining the positive directions of the X,Y, and Z axes of a coordinate system and the positive direction of rotation about an axis.

ROM—Abbreviation for read-only memory. The permanent memory of a computer that contains the computer's most fundamental operating instructions.

RPM—Abbreviation for revolutions per minute.

Save—To store data on a disk or tape.

Screen—A video display tube or CRT that displays graphic information.

Serial interface—An electrical connection that permits the linking of computers and peripherals over long distances. Also called the RS-232C interface.

Spindle speed (code word)—A code containing the letter S followed by digits. This code determines the rpm or cutting speed of the cutting spindle of the machine.

Stylus—An input device that looks like a pen and is used like a digitizer puck.

Tool function—A command that identifies a tool and calls for its selection. The address is normally a T word.

Tool length compensation—A register that eliminates the need for pre-set tooling. Allows the programmer to program all tools as if they are of equal length.

Tool offset—A correction for the tool position parallel to a controlled axis.

Turnkey—A computer system sold complete and ready to use for a specific application. You just "turn the key."

Unit—A user-defined distance, such as inches, meters, and miles.

View—A graphic representation of a two-dimensional drawing or a three-dimensional model from a specific location (viewpoint) in space.

Viewpoint—A location in three-dimensional model space from which a model is viewed.

Wireframe model—A two- or three-dimensional representation of an object consisting of boundary lines or edges of an object.

Word—A command or combination of commands that stores information upon which the machine tool acts.

Word address format—The specific group of symbols in a block of information characterized by one or more alphabetical characters that identifies the meaning of the word.

X axis—Axis of motion that is always horizontal and parallel to the workholding surface.

Y axis—Axis of motion that is perpendicular to both the X and Z axes.

Z axis—Axis of motion that is always parallel to the principal spindle of the machine.

INDEX

Page references in *italics* are illustrations.

A

Absolute positioning, 24–26, 30–31, 43
  in milling, 74, 105–106
  in turning, 188
American standard projection, 64
Arc feed. *See* Circular interpolation
Arc(s)
  incremental location of center, 72
  program examples, 137–141, *140*, 142–148
AutoCAD, 242–243
Autodesk, 242, *243*
Auto Formatting, 59
Auto Numbering, 59, 63, 67
Auxiliaries, 111, 193
Axes
  halting movement of (*see* Dwell)
  in jog mode, 63
  switching off, 111, 193
  synchronizing with spindle, 192
  X, Y, and Z, 20–21, 73
  zero, 67

B

Blocks
  format restrictions, 40–41
  independent execution, 67
  inserting extra lines in, 73
  numbering, 59, 63, 67
  skipping, 62–63, 121–122, 201–203
  start of, 37, 72
Bolt holes, 101, 306
Boring
  and canned cycles, 100
  single point tools, 175–179
  See also Counter boring

C

CAD/CAM systems, 240, 297–299
CADL, 276
Canned cycles, 44–45, 191, 192
  cancel, 74, 100

Cartesian graph, 23, *23, 28*
CBT Start Page, 53
Chatter, 47
Circular interpolation
  in milling, 78–81, 82, 142–147, *146*, 307–308
  in turning, 167–170, *168, 169,* 215–220, *219, 220*
Clamps, 116–118
Clockwise feed, 78–79
Code generation, CNC, 294–297
Color scheme, 59
Command buttons, 251–252
Comments, 122–123, 203
Computer-aided design (CAD), 240–242
Computer-aided manufacturing (CAM), 59, 243–245
Computer-integrated enterprise (CIE), 297–298
Concurrent engineering, 299
Configuration, *54,* 54–55
  reading saved, 60
Construction planes (CPLs), *249,* 249–250
Coolant
  on, 115, 198–200
  off, 111, 115–116, 193, 200
  status, 115
Coordinates
  defaulting to, 105–106
  designating, 73, 179
  displaying, 56, *105*
  in EdgeCAM, *248,* 251–252
  exercises, 304, 309
  in MasterCAM, 290, *290*
  in milling
    absolute, 24–26, 30–31, 43, 123
    incremental, 26, 31–32, 106–107, 152–158, *155*
  and program zero, 43–44
  reference points, 23
  in turning
    absolute, 188
    incremental, 188–189
Coordinate systems
  Cartesian, 27–28, 123
  in EdgeCAM, 248
  setting, *98,* 98–99, 179, 180